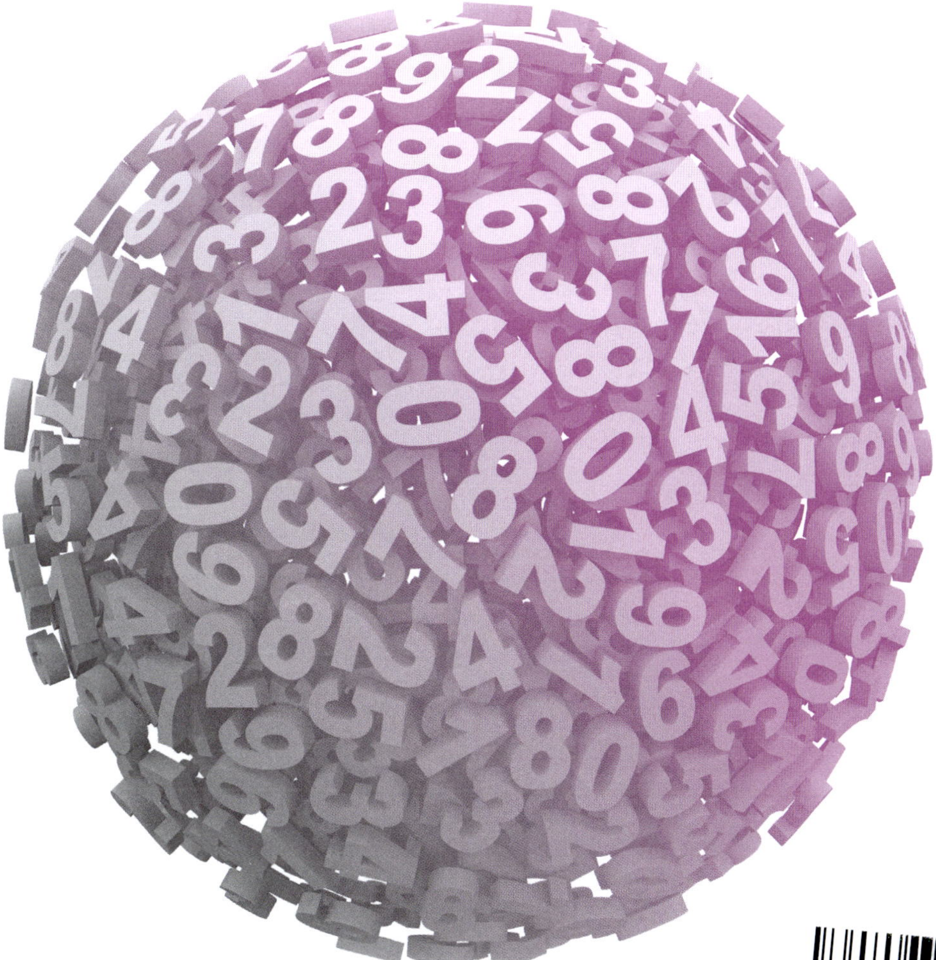

NUMBER 3/4

Walker Maths Essentials: Number 3/4
1st Edition
Charlotte Walker
Victoria Walker

Designer: Cheryl Smith, Macarn Design
Production controller: Jess Lovell

Any URLs contained in this publication were checked for currency during the production process. Note, however, that the publisher cannot vouch for the ongoing currency of URLs.

Acknowledgements
Cover photo courtesy of Shutterstock.
We wish to thank the Boards of Trustees of Darfield and Riccarton High Schools for allowing us to use materials and ideas developed while teaching. Our thanks also go to all past and present colleagues, especially Kath Wilson, who have generously shared their experience and ideas.

For product information and technology assistance,
in Australia call **1300 790 853**;
in New Zealand call **0800 449 725**

For permission to use material from this text or product, please email **aust.permissions@cengage.com**

National Library of New Zealand Cataloguing-in-Publication Data
A catalogue record for this book is available from the National Library of New Zealand.

978 0 17044733 1

Cengage Learning Australia
Level 7, 80 Dorcas Street
South Melbourne, Victoria Australia 3205

For learning solutions, visit **cengage.co.nz**

Printed in China by 1010 Printing International Limited.
4 5 6 7 24

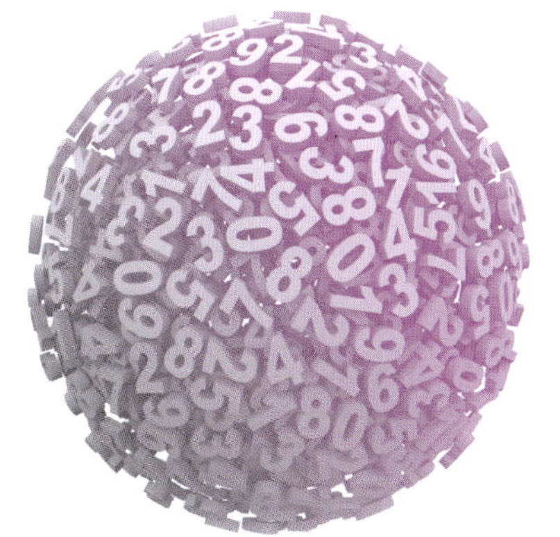

CONTENTS

Glossary

Make your own glossary of key terms:

Term	Definition	Picture/Example
BEDMAS		
Integers		
Multiple		
Factor		
Power		
Exponent		
Index		

ISBN: 9780170447331

Square root		
Numerator		
Denominator		
Mixed fraction		
Improper fraction		
Equivalent fraction		
Place value		
Decimal place		

ISBN: 9780170447331

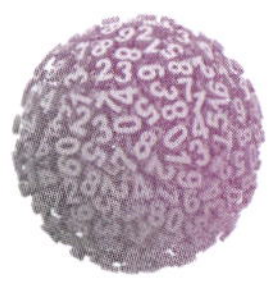

The language of mathematics

Operations

Place the terms below in the correct box.

subtract	~~add~~	multiply	fewer
total	of	take away	combine
decrease	product	out of	increase
deduct	equals	double	gives
share between	altogether	divide	minus
goes into	and	together	less
will be	plus	is	times

add

+

−

=

x

÷

 ISBN: 9780170447331

Words to operations

Write the following as calculations using one operation (+, –, x or ÷). Then write the solution to each.

	Terms	Calculation	Answer
1	Twelve **shared** between three.	12 ÷ 3	4
2	What is two **more than** five?		
3	Calculate three **times** two.		
4	Find eight **divided** by four.		
5	Nine is **added** to three.		
6	What is ten **take away** four?		
7	What is ten **shared between** two?		
8	Calculate two **multiplied by** five.		
9	What is one **increased by** seven?		
10	Find the **difference between** five and one.		
11	What is eight **plus** four?		
12	Find the **sum of** three and five.		
13	What is six **split between** three?		
14	Calculate three **less than** five.		
15	What is three **minus** one?		

ISBN: 9780170447331

Word questions

	Question	Operator (+, –, x or ÷)	Working	Answer
1	Joe has been asked to sort the tennis balls into three equal bins. There are twenty-one tennis balls. How many will be in each bin?			
2	Two friends went busking. One earned twenty-three dollars and the other got fourteen. How much did they earn in total?			
3	Nikau gets twenty dollars for each car he washes. On Saturday he washed four cars. How much did he earn?			
4	Angus and three friends shared a packet of twelve biscuits equally between them. How many biscuits did each get?			
5	Riley's mum gave him five pens to use at school. By the end of the first term, he had lost three. How many pens did he have left?			
6	Luca has ten cows and six sheep on his farm. How many cows and sheep does he have altogether?			
7	Gregory makes twelve sandwiches to share equally with his five friends. How many sandwiches does each of them get?			
8	Rosa has three ice cube trays with fourteen cubes in each. How many ice cubes does she have altogether?			
9	Natsuki has sixteen lollies. She gives nine away to her friend Lucy. How many lollies does Natsuki have left?			

ISBN: 9780170447331

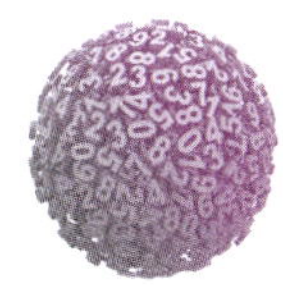

Integers

Adding and subtracting positive integers

- **–** means move **left**.
- **+** means move **right**.

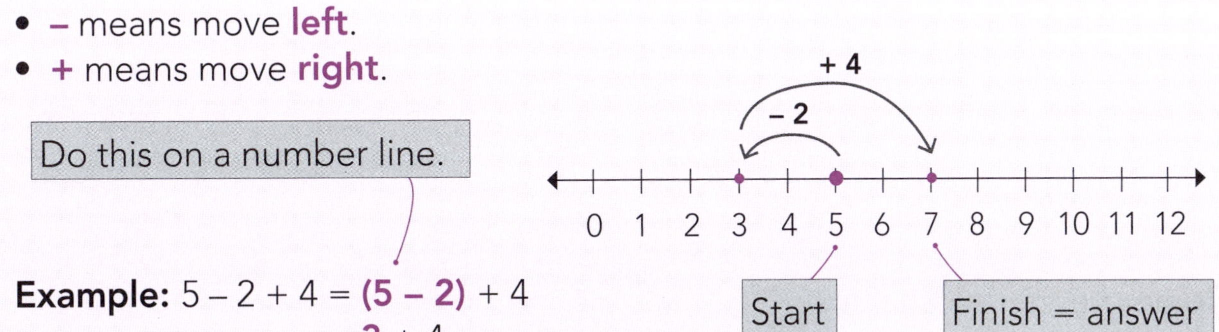

Example: $5 - 2 + 4 = \mathbf{(5 - 2)} + 4$
$= \mathbf{3} + 4$
$= 7$

Add arrows and dots to these number lines in order to complete the calculations.

1 $4 - 3 + 6 =$

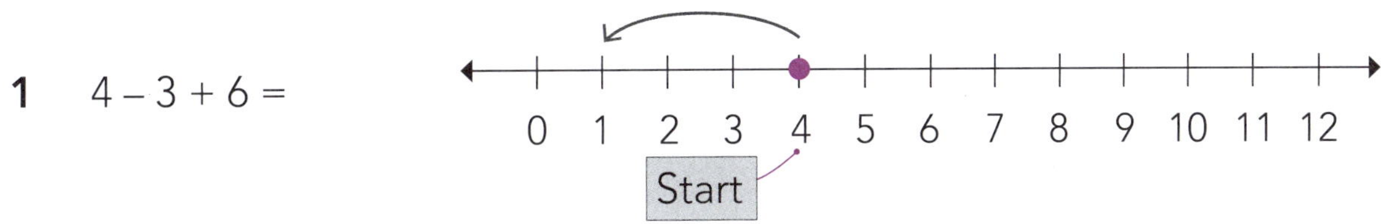

2 $6 + 3 - 7 =$

0 1 2 3 4 5 6 7 8 9 10 11 12

3 $1 + 10 - 4 =$

0 1 2 3 4 5 6 7 8 9 10 11 12

4 $5 + 7 - 10 =$

0 1 2 3 4 5 6 7 8 9 10 11 12

ISBN: 9780170447331

5 $10 - 7 - 2 + 4 =$

0 1 2 3 4 5 6 7 8 9 10 11 12

6 $6 + 3 - 7 + 1 =$

0 1 2 3 4 5 6 7 8 9 10 11 12

Complete the calculations.

7 $9 + 1 - 6 =$ ______

8 $7 - 1 + 2 =$ ______

9 $3 + 7 - 8 =$ ______

10 $4 - 1 + 8 =$ ______

11 $9 - 5 - 2 + 1 =$ ______

12 $6 + 1 - 5 - 2 =$ ______

Place these values in ascending order (smallest to largest).

13 8 12 4 2 3 10 0 6

Smallest | | | | | | | Largest

		3					

14 2 12 0 21 11 22 1 20

Smallest | Largest

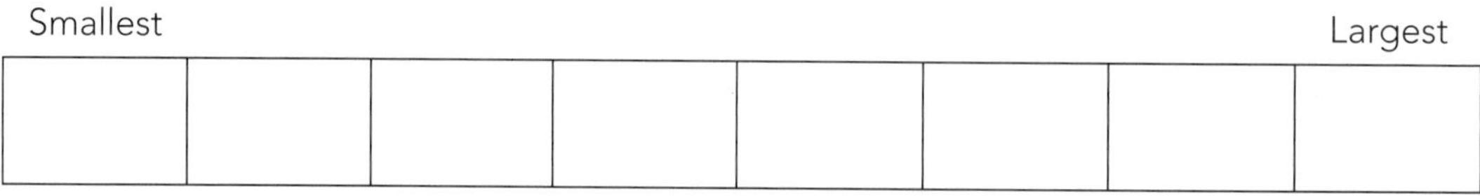

15 34 4 30 13 43 3 14 40

Smallest | Largest

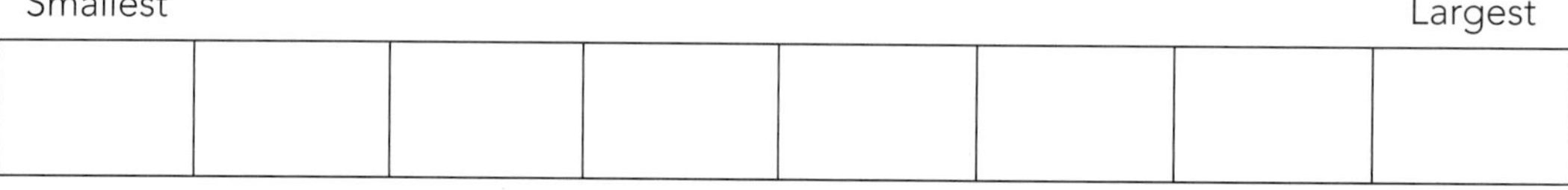

ISBN: 9780170447331

Adding and subtracting negative integers

- Negative numbers are numbers less than zero.
- On a number line they are to the left of 0.

1 Complete the number line by adding the correct values.

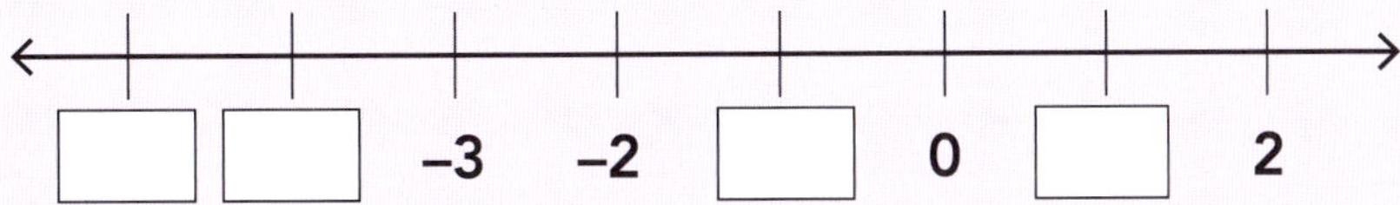

Work from left to right.

Example: –1 + 4 – 6 = **(–1 + 4)** – 6
= **3** – 6
= –3

–6
+ 4
–5 –4 –3 –2 –1 0 1 2 3 4 5 6 7
Finish
Start

Add arrows and dots to these number lines in order to complete the calculations.

2 –3 + 6 – 2 =

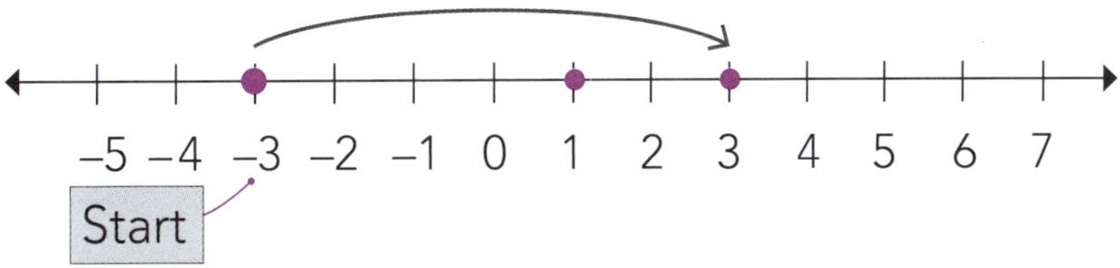

3 –4 + 2 + 7 =

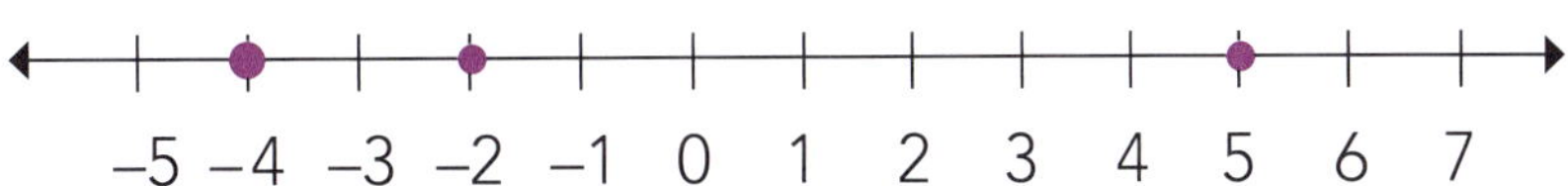

4 –2 – 3 + 8 =

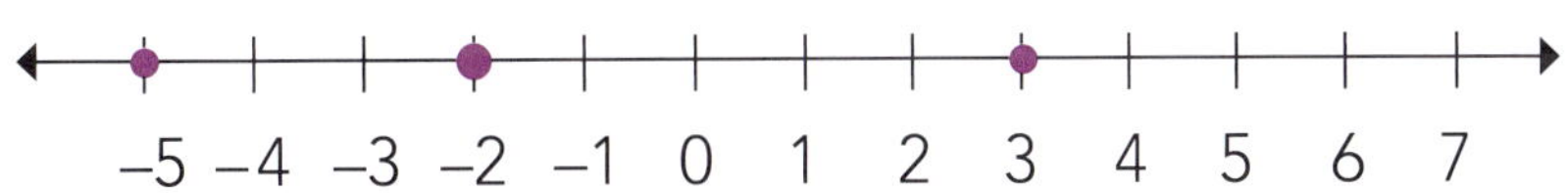

5 2 – 6 + 5 – 1 =

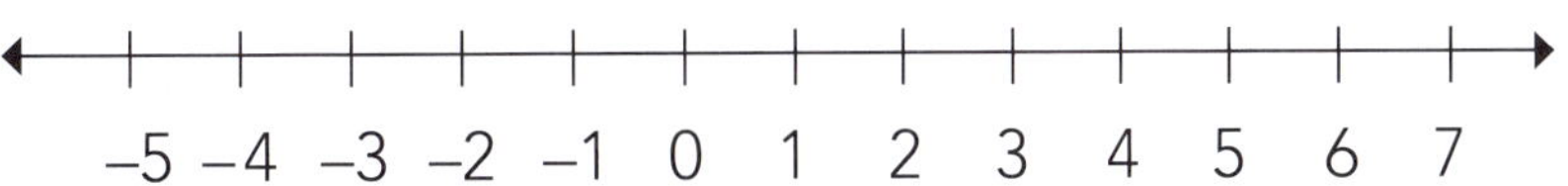

ISBN: 9780170447331

6 $-4 + 6 - 2 + 1 =$

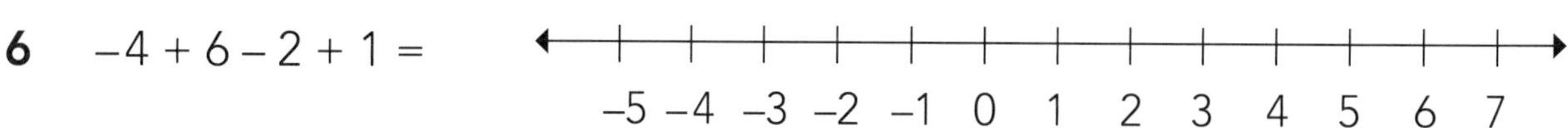

7 $5 - 7 + 3 - 4 =$

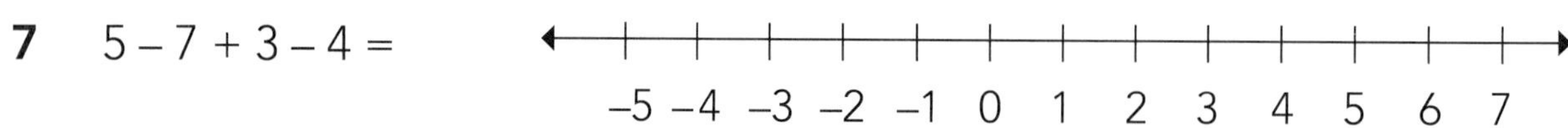

Complete the calculations.

8 $-5 + 7 =$ ________________

9 $4 - 7 + 2 =$ ________________

10 $-3 + 4 - 7 =$ ________________

11 $2 - 6 + 1 =$ ________________

12 $-5 + 4 - 2 + 6 =$ ________________

13 $6 - 4 + 1 - 5 =$ ________________

Place these values in ascending order (smallest to largest).

14 1 4 –1 0 –2 5 3 6

15 –7 12 4 0 –2 –12 7 9

Smallest							Largest

16 3 13 –3 31 0 –13 –31 1

Smallest							Largest

ISBN: 9780170447331

Integers on number lines

Here is how to work out the size of each gap between ticks on a number line.

Step 1: Calculate the **distance** between two **labelled** points.
Distance = 10 – 2 = **8**.

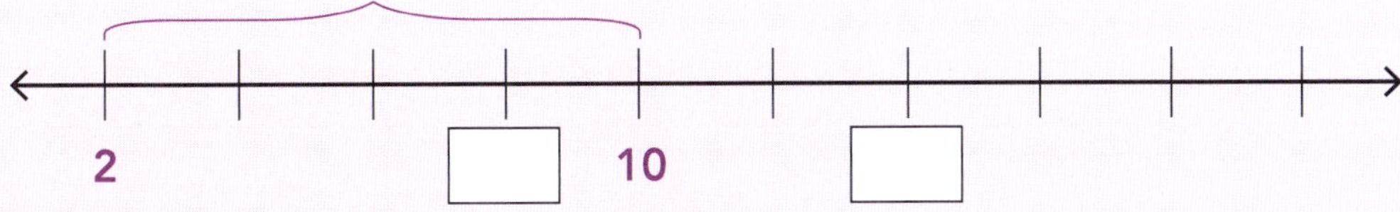

Step 2: Count the **number of gaps** between 2 and 10.
Number of gaps = 4.

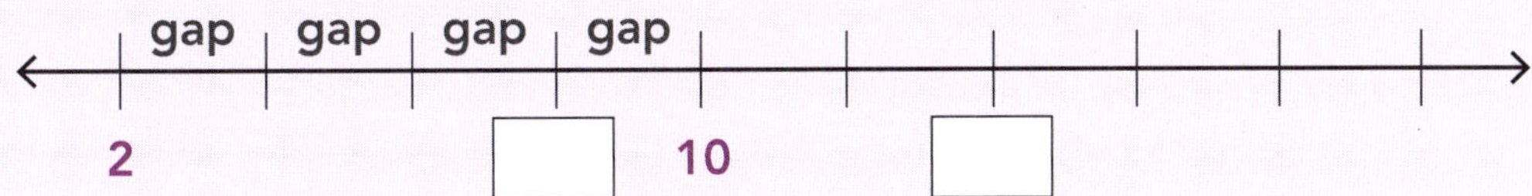

Step 3: Divide the distance by the number of gaps: $\frac{8}{4} = 2$.

Step 4: Add 2 after each gap along the number line.

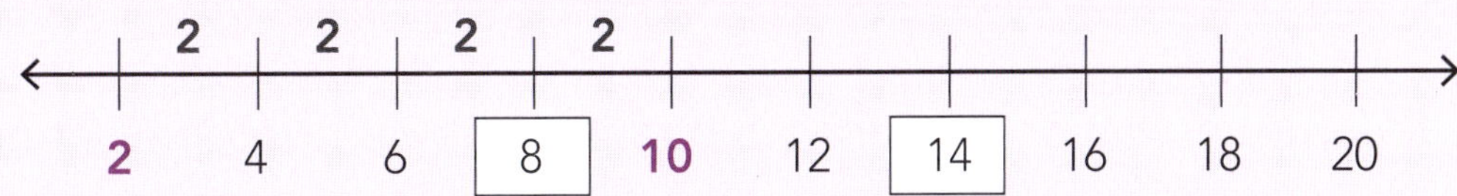

Calculate the size of each gap and write the missing integers on the number lines.

1

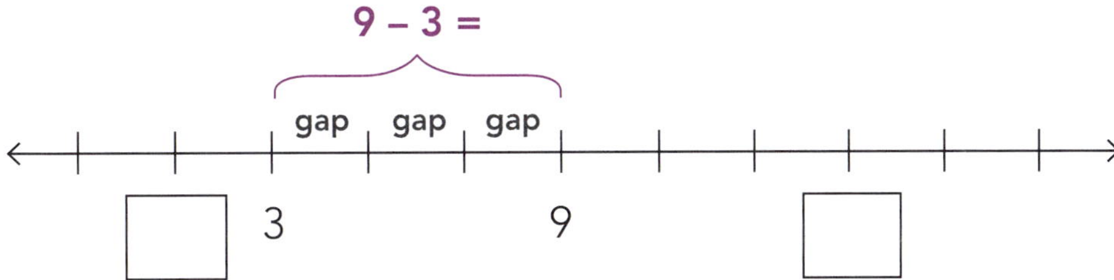

Size of gap = $\frac{\text{interval}}{\text{gaps}} = \frac{9-3}{3} = 2$.

2

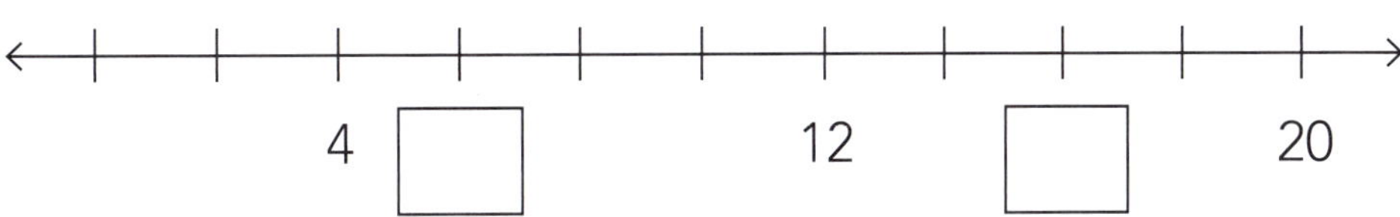

Size of gap =

3

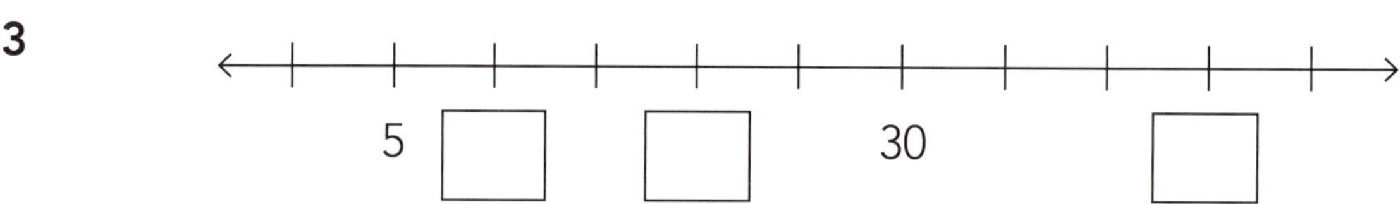

Size of gap =

4

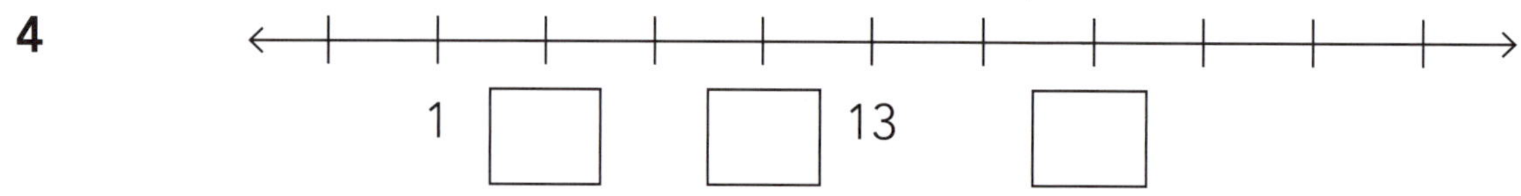

Size of gap =

5

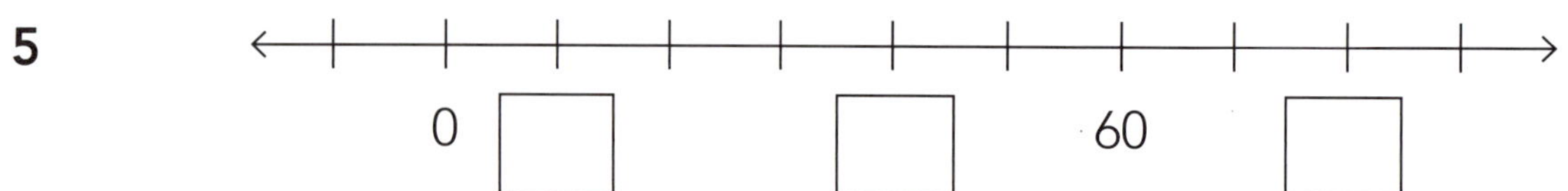

Size of gap =

6 Write the integer values for each point along the number line. Choose the most appropriate values from the list below. You will not need all the points on the list.

17	1	2	11	5	14
16	21	3	8	6	19
7	15	18	12	13	4

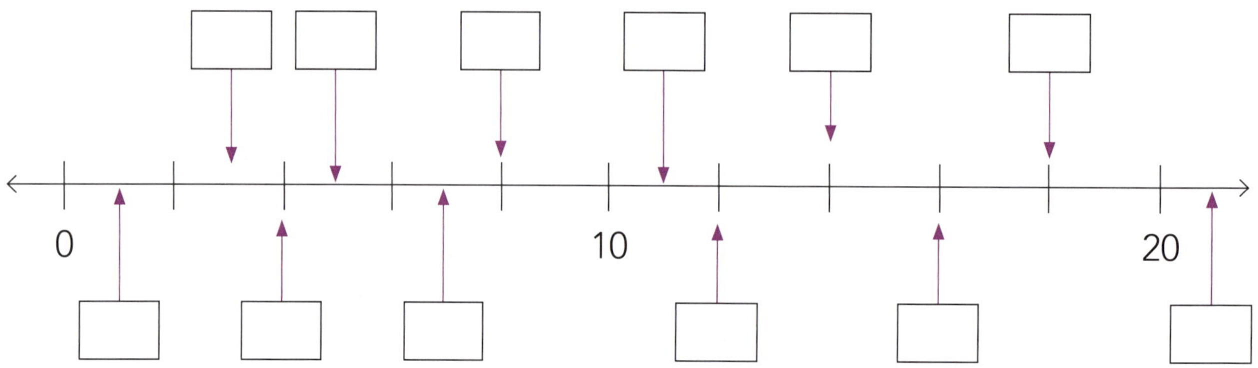

ISBN: 9780170447331

Write the missing integers on the number lines.

7

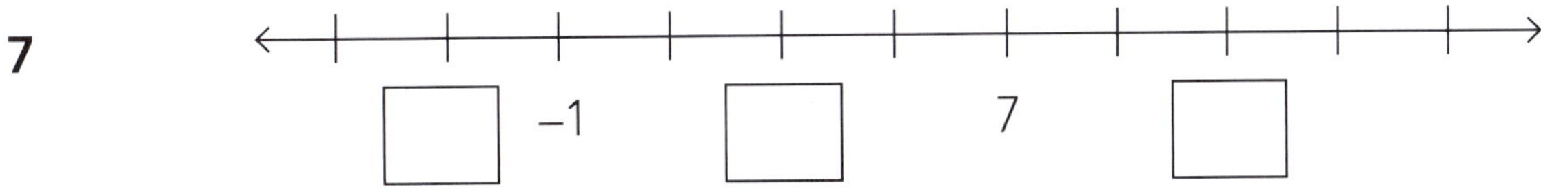

8

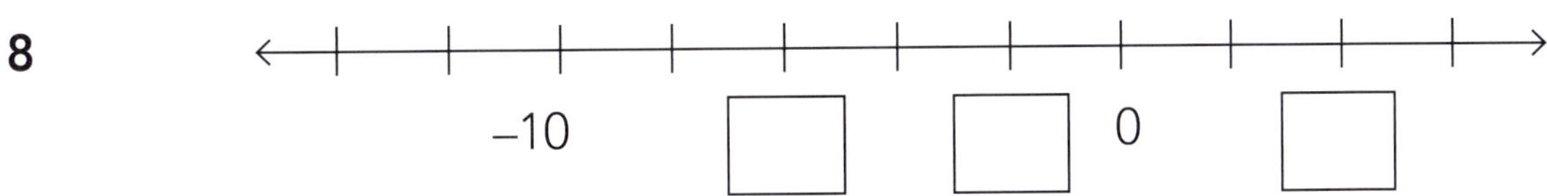

9

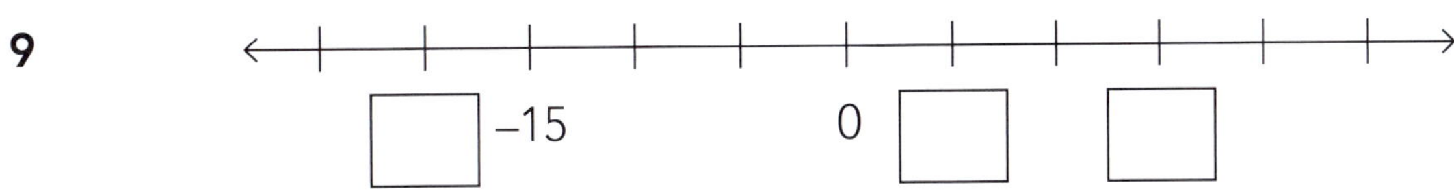

Challenge 1

Write the integer values for each point along the number line. Choose the most appropriate values from the list below. You will not need all the points on the list.

–4	–11	–1	–16	–5	–7
1	0	–10	–13	–3	–15
–9	–12	–2	–14	–17	–8

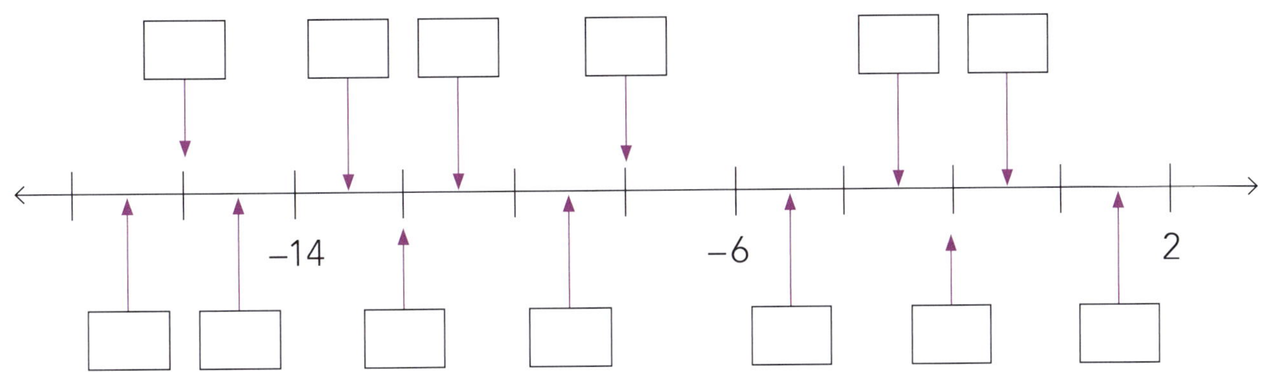

ISBN: 9780170447331

Multiplying and dividing

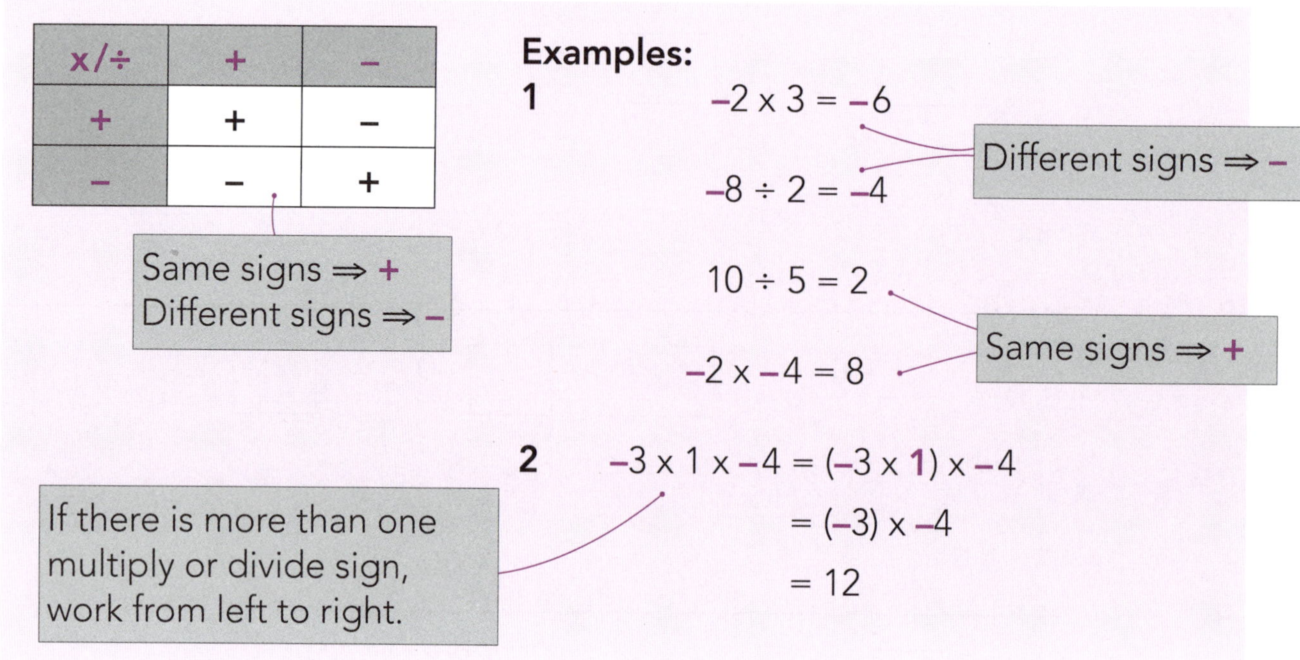

Highlight the correct answer for each of the following.

1	–2 x –4	6	–8
		8	–6
3	–5 x 1	5	–4
		4	–5
5	1 x 3 x –2	–5	6
		–6	5

2	–10 ÷ 2	–5	5
		–8	8
4	20 ÷ 4	5	16
		24	–5
6	8 ÷ –2 x 3	7	4
		12	–12

Calculate the following.

7 –3 x 2 = __________

8 –9 ÷ –3 = __________

9 –1 x –5 = __________

10 12 ÷ –2 = __________

11 8 x –2 = __________

12 –10 ÷ –2 = __________

13 6 ÷ 3 x –1 = __________

14 4 x –3 ÷ –2 = __________

ISBN: 9780170447331

Fill in the gaps in order to complete correct calculations.

15 $2 \times \square = -8$

16 $-4 \times \square = -12$

17 $\square \div -2 = 6$

18 $-4 \div \square = -1$

19 $-5 \times \square = 15$

20 $\square \div 3 = -9$

21 $-2 \times \square \times -1 = 6$

22 $6 \div \square \times -2 = -4$

23 Place the numbers into the boxes below to make true statements. You can use each number once only and you won't need them all.

6	–10	–4	4
–5	3	10	–12
2	–2	–6	–3

x  = 

÷  = 

 x  = 

ISBN: 9780170447331

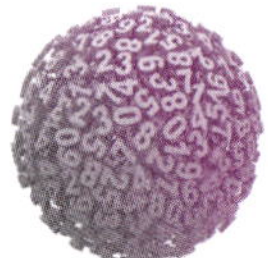

Types of numbers

Multiples

- Multiples of a number are the results of multiplying the number by another number.

Examples: The multiples of **2** are: **2**, **4**, **6**, **8**, **10**, ...
The multiples of **5** are: **5**, **10**, **15**, **20**, ...

These are the answers from the times tables.

Highlight the numbers that are multiples of the given numbers. There are two answers to each question.

1	5	20	23
		3	15
3	3	10	9
		6	19
5	10	20	100
		1	45

2	7	5	21
		13	7
4	9	18	21
		22	36
6	4	16	22
		24	18

7 Complete the table.

Number	First five multiples
2	2, 4, 6, 8, 10
5	
9	
3	
4	
10	
11	
7	
8	
6	

8 Complete the table by stating if these are true or false.

Statement	True or False
12 is a multiple of 3	
10 is a multiple of 4	
8 is a multiple of 2	
18 is a multiple of 6	
13 is a multiple of 3	
2 is a multiple of 1	
21 is a multiple of 7	
15 is a multiple of 5	
19 is a multiple of 2	
9 is a multiple of 4	

ISBN: 9780170447331

Factors

- Factors of a number are all the numbers that **divide** into it exactly.

Examples: The factors of **8** are: **1**, **2**, **4**, **8**.
The factors of **12** are: **1**, **2**, **3**, **4**, **6**, **12**.

Highlight the numbers that are factors of the given numbers. There are two answers to each question.

1	12	3	7
		4	10
3	15	8	1
		7	5
5	8	5	4
		6	2

2	10	5	6
		8	2
4	21	9	6
		7	3
6	18	8	6
		12	2

7 Complete the table.

Number	Factors
10	1, 2, 5, 10
8	
12	
9	
15	
6	
18	
21	
20	
13	
24	
50	

8 Complete the table by stating if these are true or false.

Statement	True or False
2 is a factor of 15	
1 is a factor of 7	
5 is a factor of 10	
3 is a factor of 12	
6 is a factor of 18	
2 is a factor of 1	
7 is a factor of 20	
4 is a factor of 12	
3 is a factor of 13	
4 is a factor of 20	
8 is a factor of 18	
1 is a factor of 15	

ISBN: 9780170447331

9 Highlight the numbers that belong in the boxes.

Multiples	Number	Factors
12 2 8 42 24 19 36 28	**12**	1 3 16 9 44 12 14 32 4
10 5 45 25 40 30 28 2	**10**	4 5 10 50 25 2 1 3
16 40 48 20 12 4 24 2 26	**8**	10 6 8 1 12 4 3 2 20
32 1 6 24 27 36 18 9 3	**9**	2 6 3 18 20 9 1 4 27
55 1 2 15 5 30 20 45 3	**15**	60 45 1 15 5 8 30 7 3

ISBN: 9780170447331

Square numbers

- A square number is a product of (x) two equal counting numbers (counting numbers are 1, 2, 3, …).
- Square numbers can be drawn as a square pattern of dots.

Examples: 9 **is** a square number because it is a product of 3 and 3, and it equals 3^2.

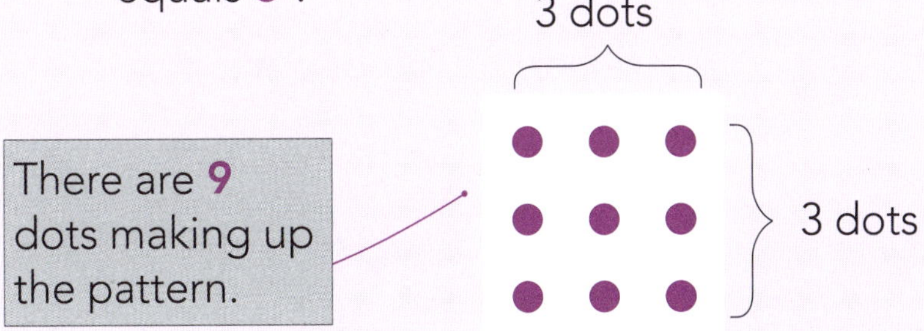

6 **is not** a square number because it cannot be written as a product of two equal numbers. Nor can 6 dots be drawn in a square.

1 Complete the patterns and fill in the missing numbers.

Square number	Picture	Notation
1		1^2
	• • • •	2^2
9		
16		4^2
	• •	
36		6^2

ISBN: 9780170447331

Powers

- Powers are used to indicate how many times a number (the base) is multiplied by itself.

The **2** is called the **base**.

The **3** is known as the **power** or the **exponent** or the **index**.

Example: $2^3 = 2 \times 2 \times 2$
$= 8$

- When a power is **2**, we say the number is **squared**, e.g. 5^2 means five **squared** (= 25).
- When a power is **3**, we say the number is **cubed**, e.g. 2^3 means two **cubed** (= 8).
- **Important:** **anything^1 = itself** e.g. $7^1 = 7$
  anything^0 = 1 e.g. $7^0 = 1$

The power indicates how many times you need to multiply.

Examples: $4^3 = 4 \times 4 \times 4$
$= 64$

$3^2 = 3 \times 3$
$= 9$

Finding powers on your calculator

- Powers of numbers can get really big, so knowing how to find them on your calculator is **very** useful.

For squares: use a button that looks like this:

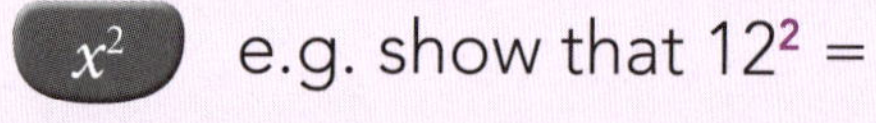

e.g. show that $12^2 = 144$.

For cubes: use a button that looks like this:

x^3 e.g. show that $6^3 = 216$.

For all other powers: use a button that looks like this:

 ISBN: 9780170447331

Write the following as powers.

1 $4 \times 4 =$ ______

2 $6 \times 6 \times 6 =$ ______

3 $5 \times 5 \times 5 \times 5 =$ ______

4 $9 \times 9 =$ ______

5 $8 =$ ______

6 $3 \times 3 \times 3 =$ ______

7 $2 \times 2 \times 2 \times 2 \times 2 =$ ______

8 $1 \times 1 \times 1 =$ ______

Write out what the following mean and then calculate their values.

9 $2^3 = 2 \times 2 \times 2 =$ ______

10 $3^3 =$ ______

11 $6^2 =$ ______

12 $8^2 =$ ______

13 $7^3 =$ ______

14 $4^1 =$ ______

15 $5^3 =$ ______

16 $10^3 =$ ______

17 $13^2 =$ ______

18 $20^3 =$ ______

Use your calculator to find the values of the following.

19 $7^2 + 4^3 =$ ______

20 $3^2 - 2^3 =$ ______

21 $4^2 \times 5^2 =$ ______

22 $10^2 \div 2^2 =$ ______

23 $\frac{4^3}{2^3} =$ ______

24 $5^2 + 4^3 - 3^2 =$ ______

ISBN: 9780170447331

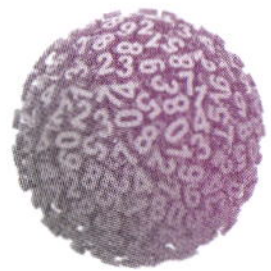

Roots

- Finding a root is the opposite of finding a power.
- The **square root** is written as $\sqrt{\ }$. You do not need to write $\sqrt[2]{\ }$.

  e.g. $\sqrt{9} = \sqrt{3 \times 3}$
  $= 3$
- The **cube root** is written as $\sqrt[3]{\ }$.

  e.g. $\sqrt[3]{8} = \sqrt[3]{2 \times 2 \times 2}$
  $= 2$

You can do this on your calculator with either 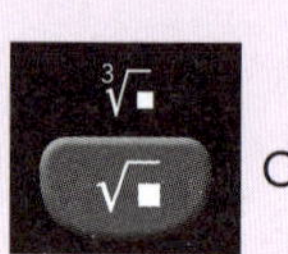 or 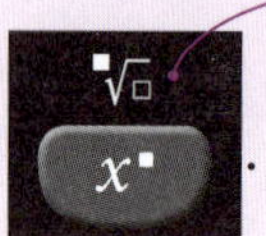.

To access the yellow function, use the shift key.

Find the following roots.

1 $\sqrt{4} =$ ______ 2 $\sqrt{25} =$ ______

3 $\sqrt{9} =$ ______ 4 $\sqrt{16} =$ ______

5 $\sqrt{36} =$ ______ 6 $\sqrt{81} =$ ______

7 $\sqrt{100} =$ ______ 8 $\sqrt{49} =$ ______

9 $\sqrt[3]{8} =$ ______ 10 $\sqrt[3]{27} =$ ______

11 $\sqrt[3]{125} =$ ______ 12 $\sqrt[3]{512} =$ ______

13 $\sqrt[3]{216} =$ ______ 14 $\sqrt[3]{64} =$ ______

ISBN: 9780170447331

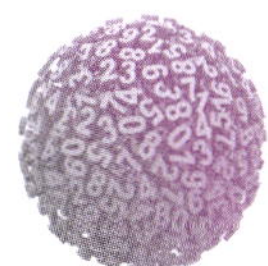

Order of operations

- **BEDMAS** helps us to remember the order of operations.
- If you have several division signs or several multiplication signs, work from left to right.

Fill in the table below and use it to help you remember the order to use in calculations:

B	
E	
D	
M	
A	
S	

Remember: when there is more than one of these, work from left to right.

Examples:

1 $4 + 8 \times 2$

$= 4 + (8 \times 2)$ — Multiplication first

$= 4 + 16$ — Addition last

$= 20$

2 $1 + (6 - 2) \div 2$ — Brackets first

$= 1 + 4 \div 2$ — Division next

$= 1 + 2$ — Addition last

$= 3$

Highlight the correct answer for each of the following.

1	$8 - 2 \times 2$	4	8
		2	12
3	$3 + 2^2$	5	7
		–1	1
5	$(6 - 1)^2$	36	25
		5	1
7	$12 \div (3 + 1)$	4	3
		8	5

2	$\frac{(2 + 4)}{2}$	5	3
		2	4
4	$2 \times (7 - 1)$	8	6
		12	13
6	$8 + 6 \div 2$	12	10
		7	11
8	$4 - \frac{(3 + 3)}{6}$	1	4
		6	3

ISBN: 9780170447331

Using BEDMAS, calculate the following.

9 $5 + 3 \times 2 =$ ____________

10 $(2 + 1) \times 3 =$ ____________

11 $1 + 3^2 =$ ____________

12 $9 - 6 \div 3 =$ ____________

13 $(2 + 2)^2 =$ ____________

14 $10 \div 5 + 2 =$ ____________

15 $\frac{(6 + 4)}{2} =$ ____________

16 $9 - 3 \times 2 - 2 =$ ____________

17 $(4 + 5 \times 1) \div 3 =$ ____________

18 $\frac{21}{(3 + 4)} =$ ____________

19 $2^2 \times 3 + 1 =$ ____________

20 $20 - 3^2 \times 2 =$ ____________

21 $4 + \frac{9}{3} =$ ____________

22 $8 - 3 \times 2 + 2^2 =$ ____________

23 Fill the gaps with operations so that the following calculations are correct. You may use only +, – and x operations, and you must use one of each in every question.

4 + 3 – 2 x 1 = 5	4 3 2 1 = 13
4 3 2 1 = 3	4 3 2 1 = 11
4 3 2 1 = 9	4 3 2 1 = –1

 ISBN: 9780170447331

	Question	Working	Answer
24	Start with twelve, **add** three, **subtract** two and **multiply** the answer by one.	$(12 + 3 - 2) \times 1$	13
25	Ten is **added** to three, then four is **subtracted**, **multiply** the result by two.	3 10 4 2	
26	**Multiply** three by four, then **increase** this by eight.		
27	Take eight, **share** it between two and **subtract** one.		
28	Three is **subtracted** from nine, the result is **multiplied** by two.		
29	Start with eleven, **take away** five, then **divide** the answer by two.		
30	Take one, then **times** by nine, **split** the result in three.		
31	Twelve is **divided** by three, the result is **multiplied** by two, then add one.		
32	Find the **difference** between ten and six. **Multiply** the result by three and **subtract** two.		
33	**Split** sixteen into four. **Divide** the result by two, then **add** twelve.		

ISBN: 9780170447331

Using your calculator

Do the calculations below using your calculator, then turn your calculator upside down and add the word to the story. Be careful to check that it makes sense.

Example: $195^2 + 51 = 38076$, upside down this spells GLOBE.

A trip to the zoo

1 ____________ and 2 ____________ were at a 3 ____________. They were at the zoo but couldn't find 4 ____________ anywhere. They were supposed to meet her back by the 5 ____________ but they couldn't 6 ____________ her.
7 ____________ came round the corner and shook his head — still no sign of 8 ____________. They couldn't 9 ____________. They were scared. They felt 10 ____________. Then 11 ____________ heard a familiar 12 ____________.
13 ____________ came trotting round the corner. 14 ____________ let out a 15 ____________ and shouted at 16 ____________. Thank goodness, they wouldn't have to tell Mum.

1 $10^2 \times \sqrt{9} + \sqrt{4}$

2 $(22^2 + 6^2) \times (24^2 + 5^2 + 10) - \sqrt{9}$

3 $\sqrt{49} + 11 \times \sqrt{25} \times 10^2$

4 $\sqrt[3]{8} \times (6^3 + 60) \times 12 \times 53 + 1$

5 $9^2 + 5^2 \times ((10^2 + 7) \times 200 + 4)$

6 $(\sqrt[3]{216} \times 11 + 1) \times \sqrt[3]{125}$

7 $(2^3 + 1) \times \frac{25}{2.5} \times (18^2 + 33 - \sqrt{16})$

8 $24^4 + 23(30^2 - 50 - \sqrt{121})$

9 $5^2 \times \frac{126}{\sqrt{81}} - \frac{99}{\sqrt{9}}$

10 $10^3 - (\sqrt[3]{125} \times \frac{230}{\sqrt{25}} - 1)$

11 $(4^2 - 1) \times (5^2 - \sqrt{25}) + \sqrt[3]{8}$

12 $\frac{64^2 + 112}{\sqrt{16}} \times (330 + 5^2 + \sqrt[3]{27})$

13 $(\frac{38}{2} \times 30 - 1) \times (\sqrt{100} \times 62 - 3)$

14 $\frac{50^2}{2} - (\sqrt{144} \times (13 \times \sqrt{36} + 1))$

15 $(5^2 - \sqrt{4}) \times \sqrt[3]{42\,875}$

16 $\frac{1\,230}{(1 + 2)} \times (9^3 + 46) - (\sqrt{9} \times 11)$

 ISBN: 9780170447331

Fractions

The language of fractions

Complete the table.

Fraction	Word	Meaning
$\frac{1}{2}$	half	÷ 2
	quarter	
		÷ 3
$\frac{1}{5}$		
	sixth	
		÷ 7
$\frac{1}{8}$		
	ninth	

ISBN: 9780170447331

Shading fractions

- Fractions are a way of writing numbers or parts of numbers.
- The **bottom** number of a fraction is how many parts there are, the **top** number is how many of these parts are shaded.

Examples:

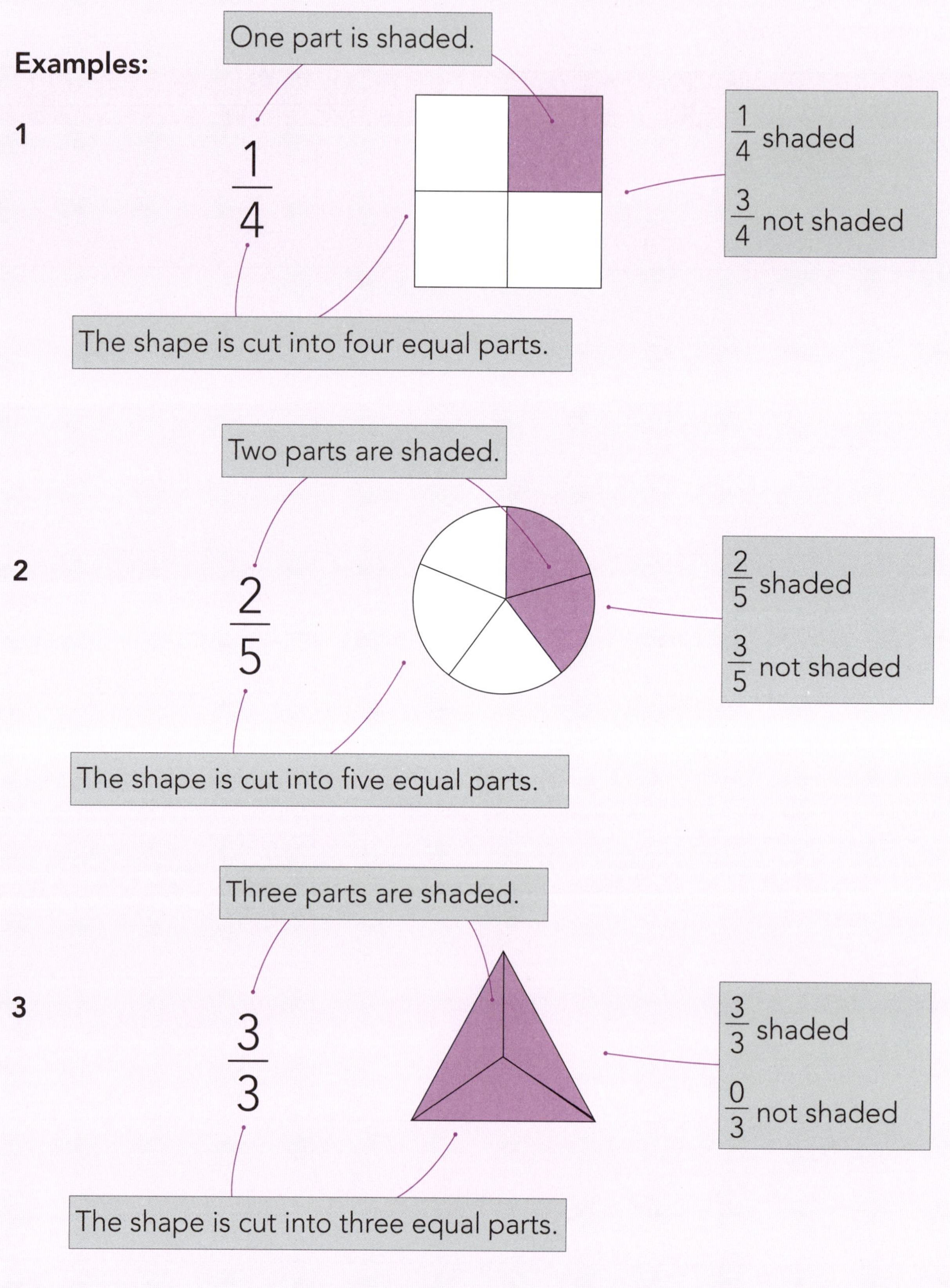

ISBN: 9780170447331

What fractions of these diagrams are shaded and not shaded?

1

Shaded $\frac{}{3}$

Not shaded $\frac{}{3}$

2

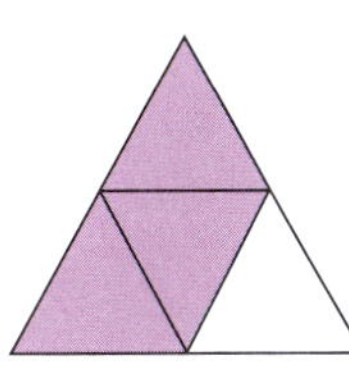

Shaded $\frac{3}{}$

Not shaded $\frac{1}{}$

3

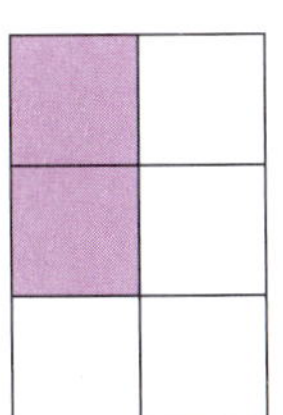

Shaded $\frac{}{6}$

Not shaded $\frac{}{6}$

4

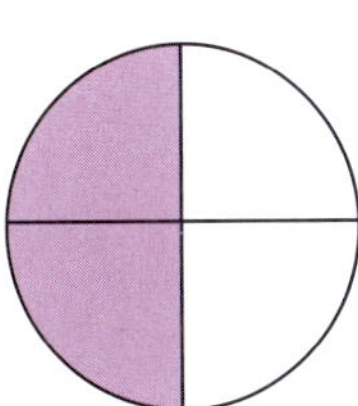

Shaded ______

Not shaded ______

5

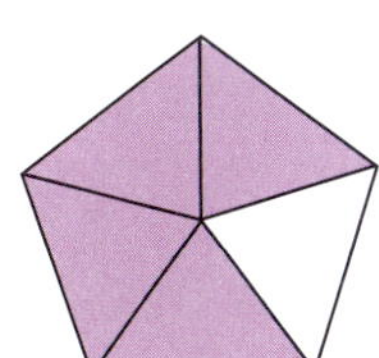

Shaded ______

Not shaded ______

6

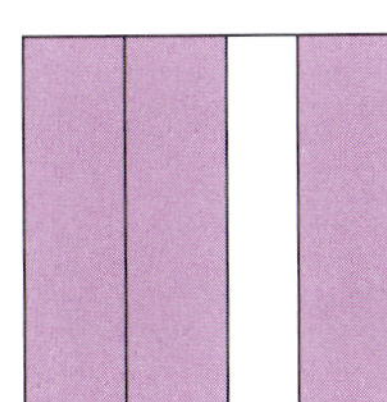

Shaded ______

Not shaded ______

7

Shaded ______

Not shaded ______

8

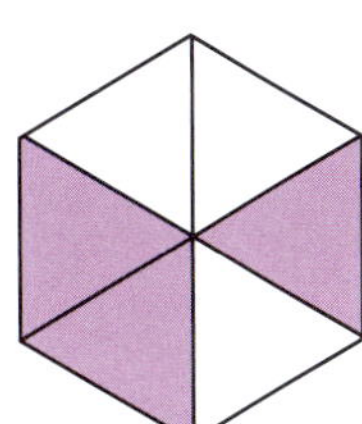

Shaded ______

Not shaded ______

9

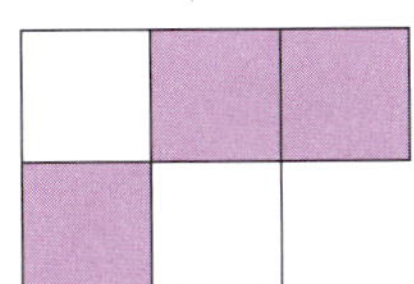

Shaded ______

Not shaded ______

10

Shaded ______

Not shaded ______

11

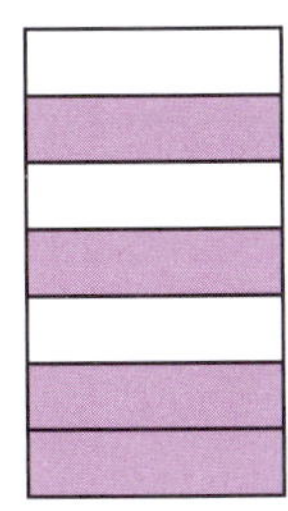

Shaded ______

Not shaded ______

12

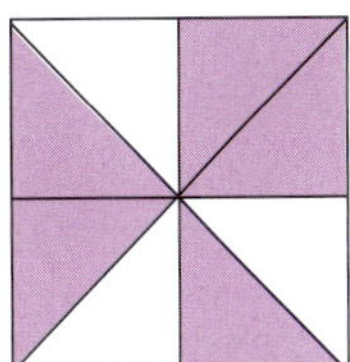

Shaded ______

Not shaded ______

ISBN: 9780170447331

Shade the diagram to match the fraction.

13

$\frac{3}{4}$

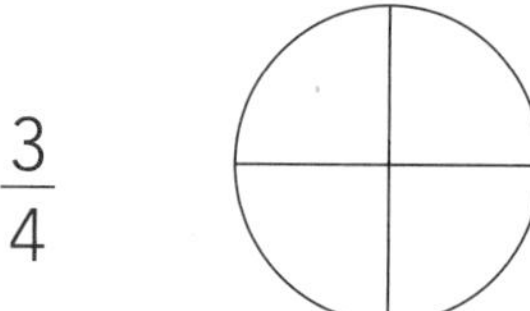

14

$\frac{5}{6}$

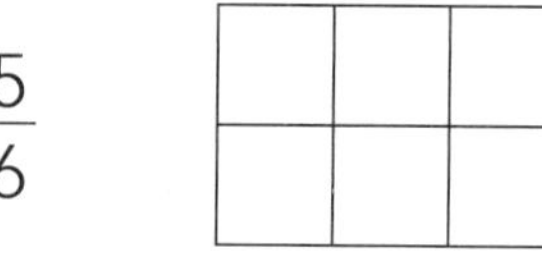

15

$\frac{2}{5}$

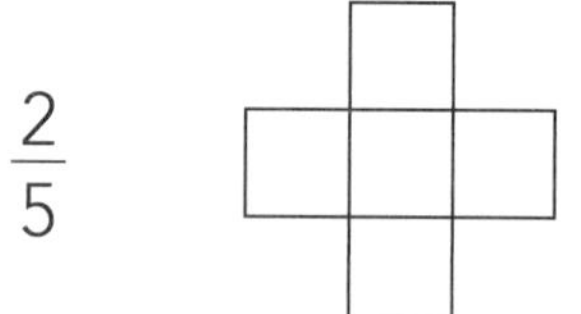

16

$\frac{2}{3}$

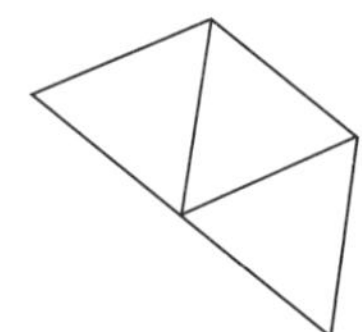

17

$\frac{1}{4}$

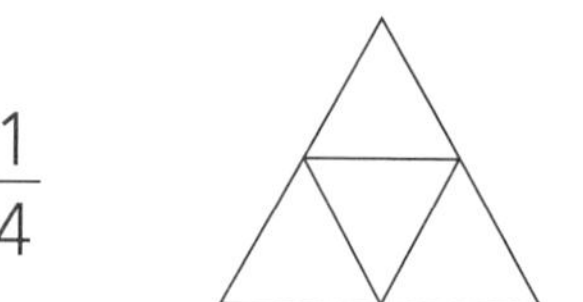

18

$\frac{3}{7}$

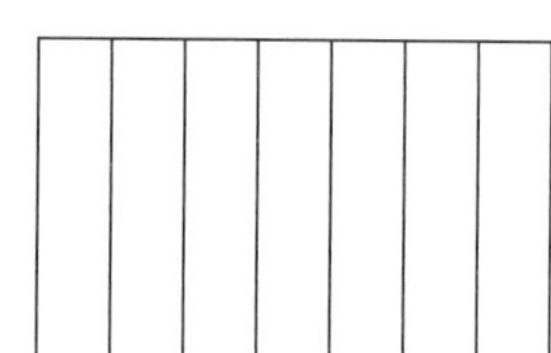

19

$\frac{5}{8}$

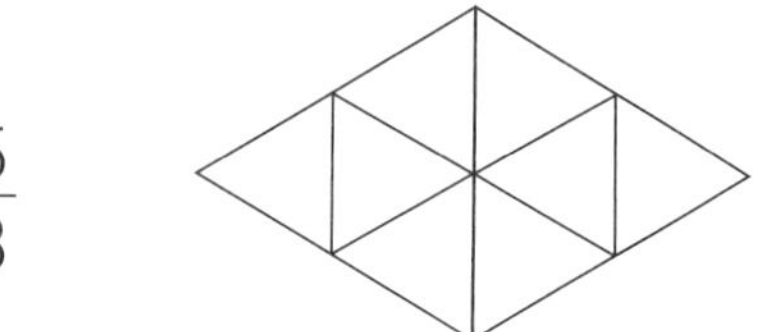

20

$\frac{5}{9}$

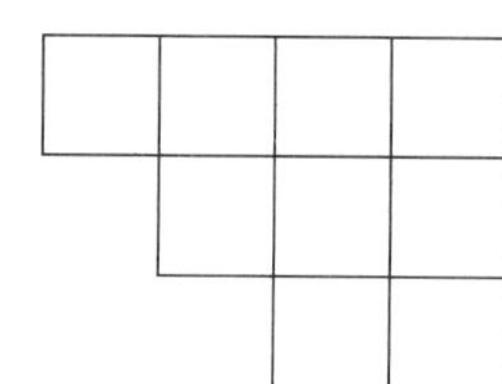

21

$\frac{4}{5}$

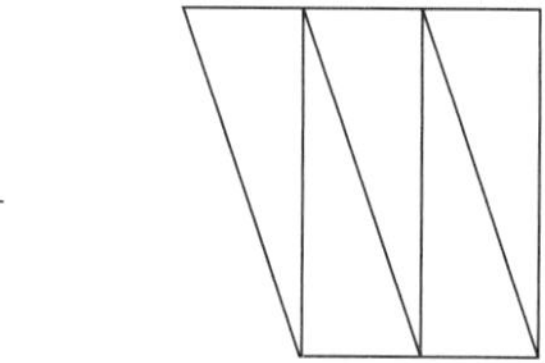

22

$\frac{7}{8}$

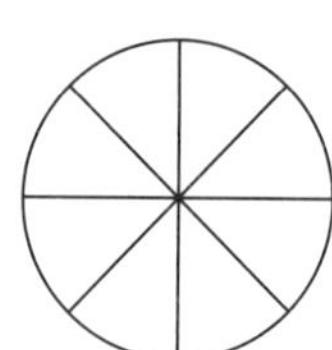

23

$\frac{3}{10}$

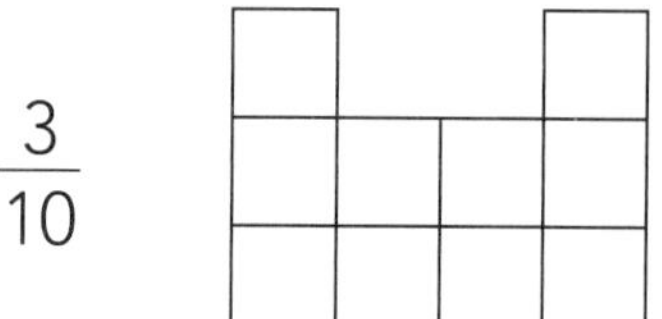

24

$\frac{1}{2}$

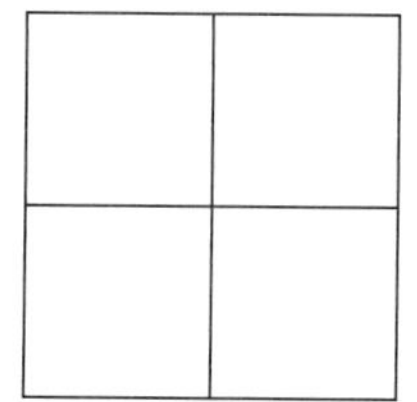

ISBN: 9780170447331

Numerators and denominators

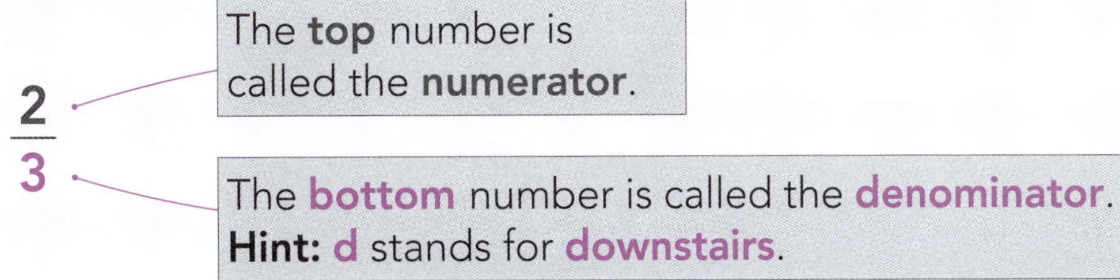

Numerators

1 Shade the bars as indicated.

Fraction

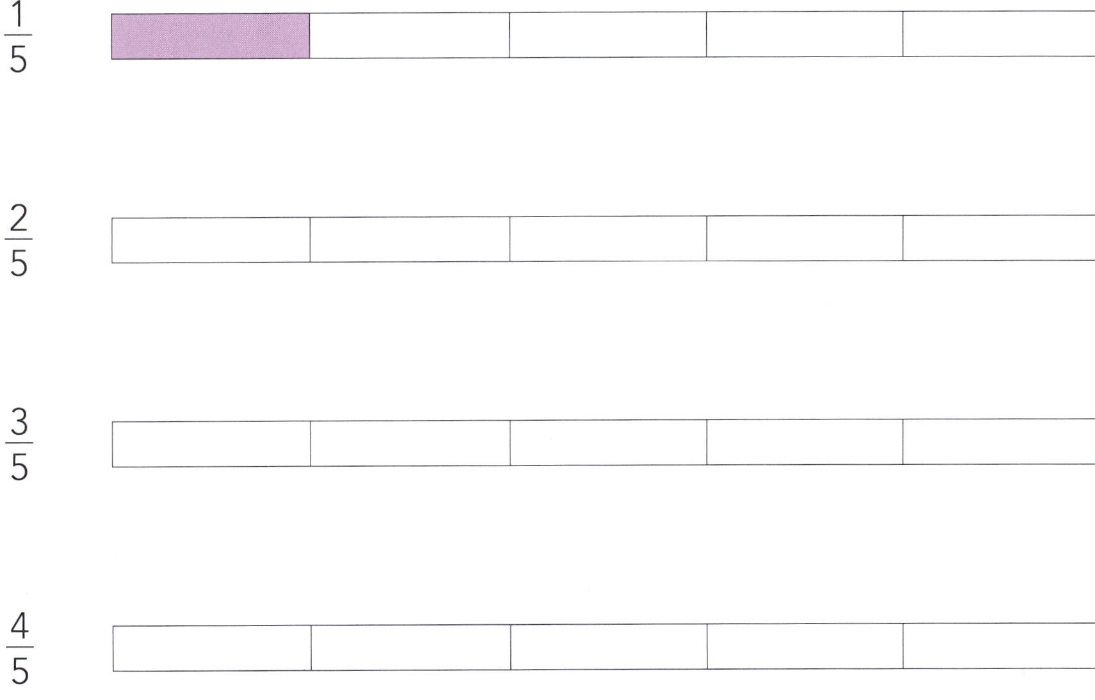

2 Cross out the word to make a true statement:

When the numerator gets bigger, we shade **more/less** of the bar.

Complete the fractions so they match the diagrams, then highlight the larger one.

3

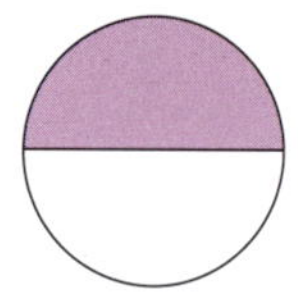

$\frac{}{2}$ $\frac{}{2}$

4

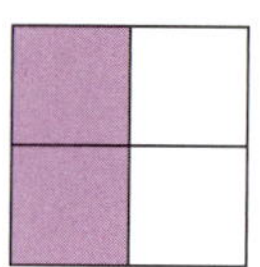

$\frac{}{4}$ $\frac{}{4}$

5

$\frac{}{5}$ $\frac{}{5}$

6

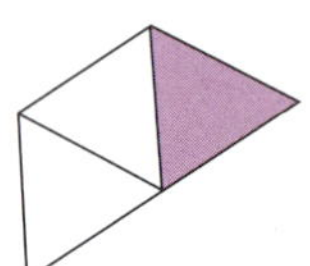

— —

7

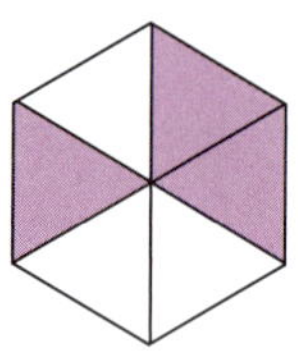

— —

8

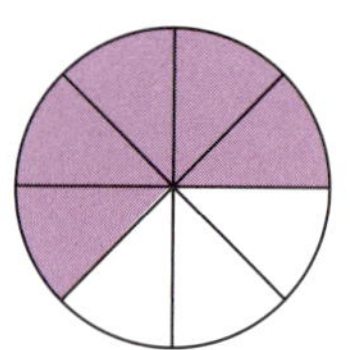

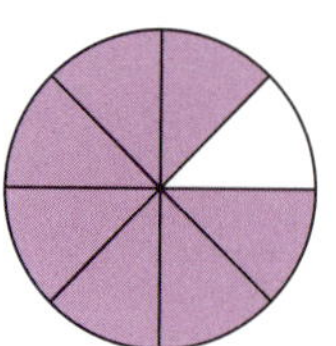

— —

Circle or highlight which fraction is larger.

9 $\frac{4}{5}$ $\frac{2}{5}$

10 $\frac{2}{7}$ $\frac{5}{7}$

11 $\frac{1}{4}$ $\frac{4}{4}$

12 $\frac{7}{10}$ $\frac{5}{10}$

13 $\frac{8}{8}$ $\frac{7}{8}$

14 $\frac{1}{2}$ $\frac{3}{4}$

 ISBN: 9780170447331

Denominators

15 Shade the bars as indicated.

Fraction	
$\frac{1}{2}$	
$\frac{1}{3}$	
$\frac{1}{4}$	
$\frac{1}{5}$	
$\frac{1}{6}$	
$\frac{1}{7}$	
$\frac{1}{8}$	

16 Cross out the word to make a true statement:

When the denominator gets bigger, we shade **more/less** of the bar.

ISBN: 9780170447331

Complete the fractions so they match the diagrams, then highlight the larger one.

17

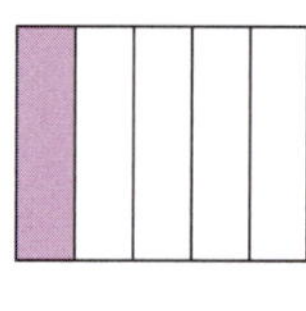

$\frac{1}{\quad}$

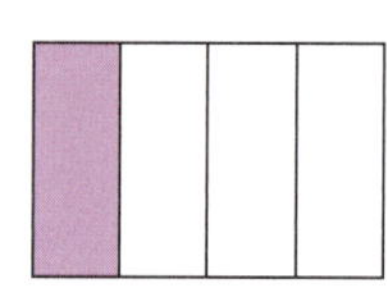

$\frac{1}{\quad}$

18

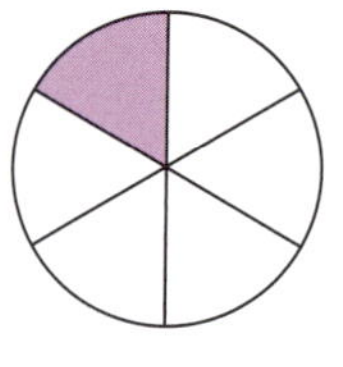

$\frac{1}{\quad}$

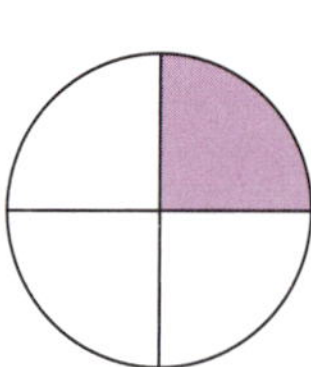

$\frac{1}{\quad}$

19

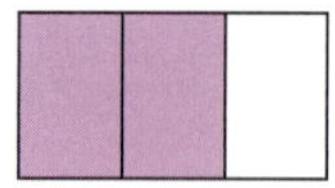

$\frac{2}{\quad}$

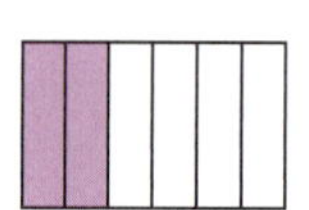

$\frac{2}{\quad}$

20

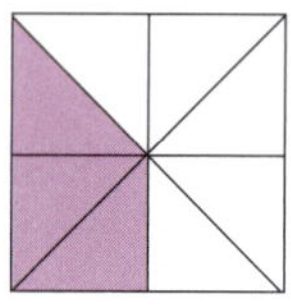

$\frac{\quad}{\quad}$

$\frac{\quad}{\quad}$

21

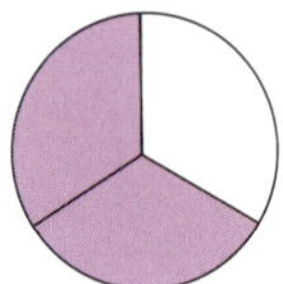

$\frac{\quad}{\quad}$ $\frac{\quad}{\quad}$

22

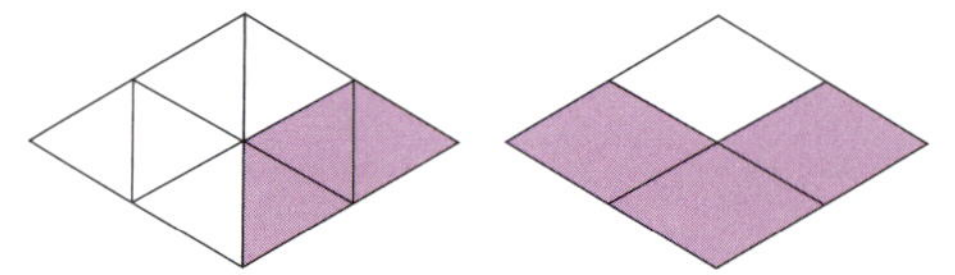

$\frac{\quad}{\quad}$ $\frac{\quad}{\quad}$

Circle or highlight which fraction is larger.

23 $\frac{1}{3}$ $\frac{1}{4}$

24 $\frac{5}{6}$ $\frac{5}{8}$

25 $\frac{2}{5}$ $\frac{2}{7}$

26 $\frac{4}{7}$ $\frac{4}{10}$

27 $\frac{1}{8}$ $\frac{1}{6}$

28 $\frac{2}{2}$ $\frac{2}{3}$

ISBN: 9780170447331

29 **a** Shade the diagrams so they match the fractions.

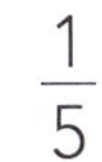 $\frac{3}{5}$  $\frac{5}{5}$ $\frac{4}{5}$

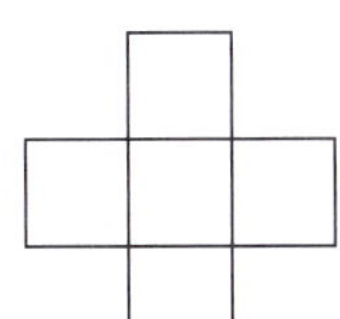 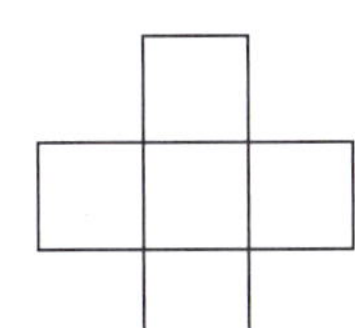 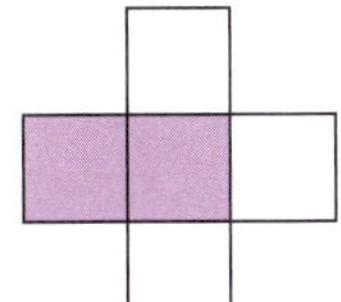 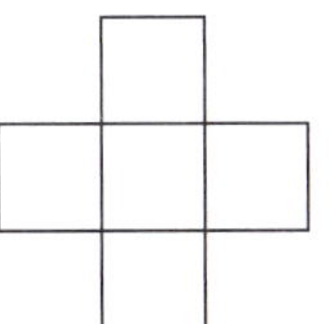 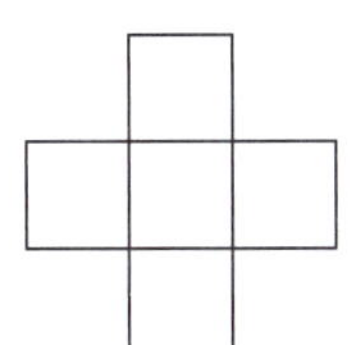

b Write the fractions in ascending order.

Smallest						Largest
		$\frac{2}{5}$				

30 **a** Shade the diagrams so they match the fractions.

$\frac{1}{6}$ $\frac{1}{3}$ $\frac{1}{2}$ $\frac{1}{4}$ $\frac{1}{5}$

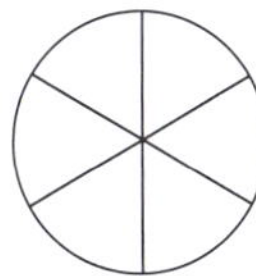 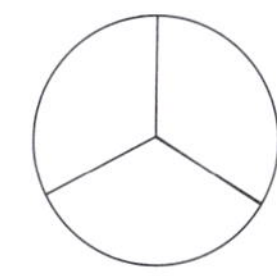 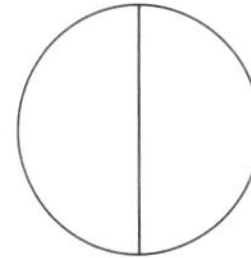 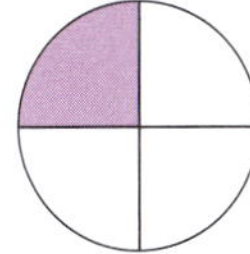 

b Write the fractions in ascending order.

Smallest						Largest
			$\frac{1}{4}$			

What's my fraction?

Circle or highlight the fraction that best fits all of the clues.

My denominator has two digits.
My denominator is an odd number.
My numerator is the square of two.

$\frac{5}{7}$	$\frac{4}{11}$	$\frac{4}{10}$
$\frac{2}{17}$	$\frac{1}{3}$	$\frac{8}{13}$

ISBN: 9780170447331

Equivalent fractions

- Equivalent fractions are fractions that are written with different denominators, but which have the same value.

$\frac{1}{2}$

$\frac{2}{4}$

Shading $\frac{1}{2}$ of a shape is the same as shading $\frac{2}{4}$.

Write the fractions that match the pictures.

1

$\frac{\ }{6}$ 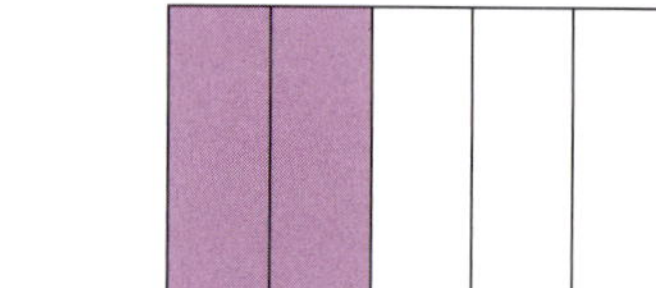

$\frac{\ }{3}$ 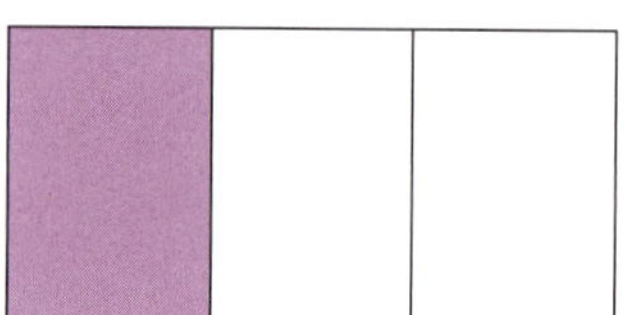

2

$\frac{\ }{8}$ 

$\frac{\ }{4}$ 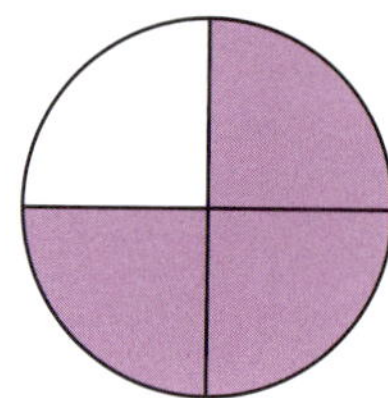

3

$\frac{\ }{3}$ 

$\frac{\ }{9}$ 

4

$\frac{\ }{10}$ 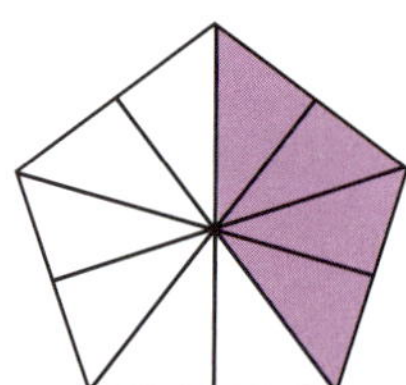

$\frac{\ }{5}$ 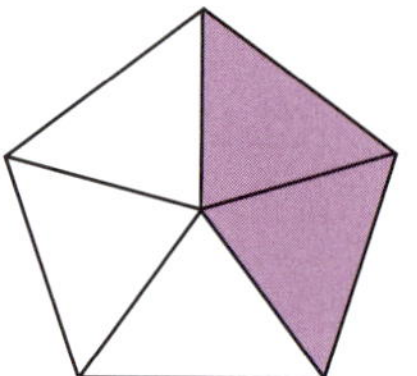

 ISBN: 9780170447331

Simplifying fractions

- Usually we want our answers in the simplest possible form.
- This means writing the fraction with the lowest possible whole numbers.
- Remember, what you do to the numerator, you must also do to the denominator.

Example:

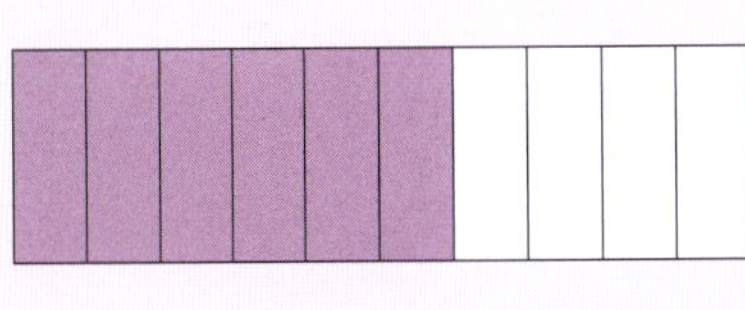

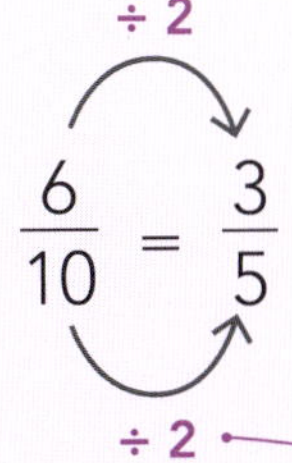

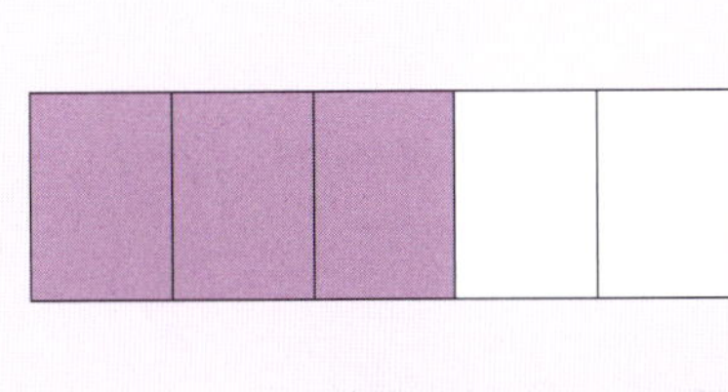

- If the numbers aren't divisible by 2, then try 3 or 5.
- Keep going until you can't go any further.

What you do to the numerator, you do to the denominator.

Fully simplify these fractions.

1

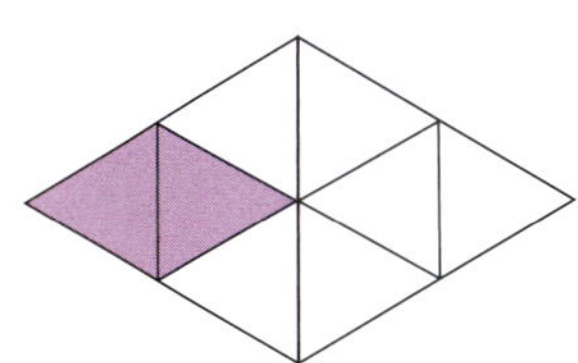

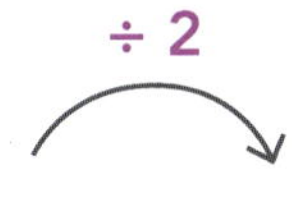

$\frac{}{8} = \frac{}{4}$

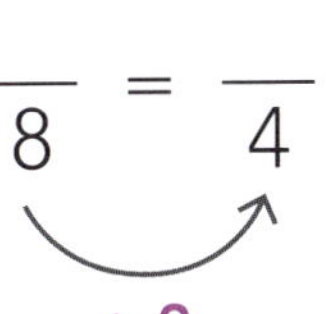

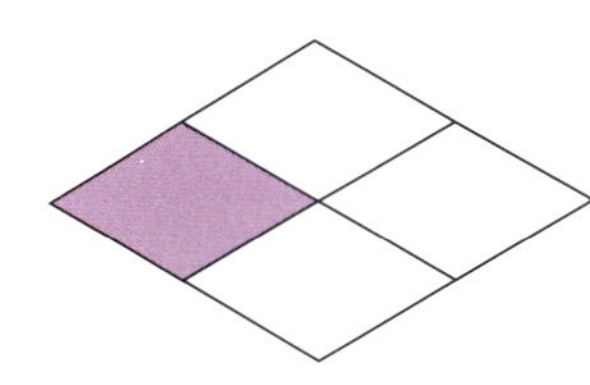

2

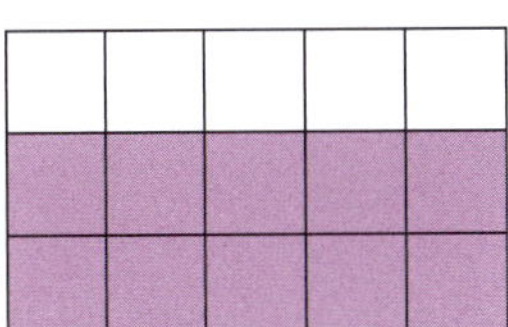

÷ 5

$\frac{}{15} = \frac{}{3}$

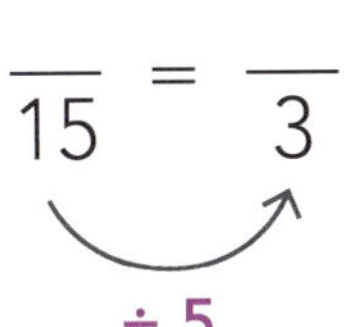

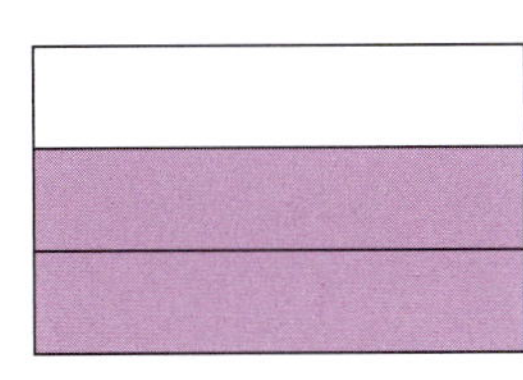

3

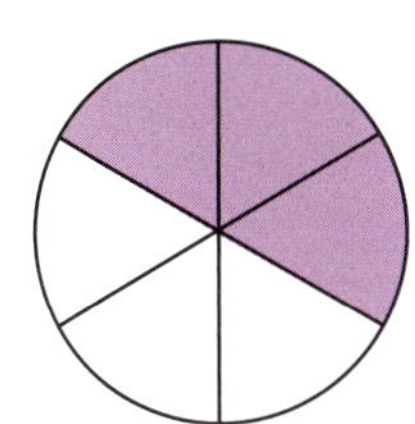

÷ 3

$\frac{}{6} = \frac{}{}$

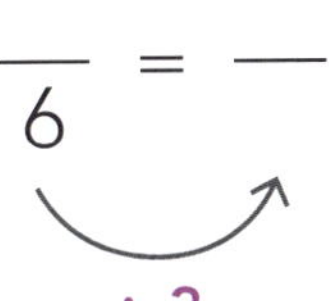

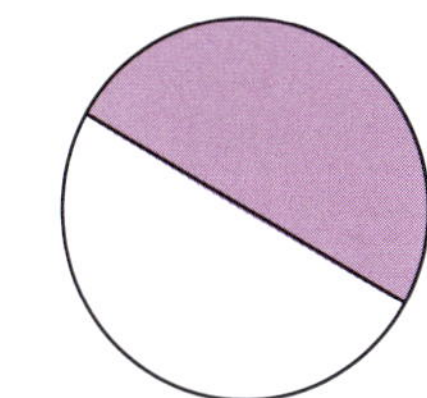

ISBN: 9780170447331

Fill in the gaps to create equivalent fractions.

4

÷ 2

$$\frac{6}{8} = \frac{}{}$$

÷ 2

5

÷ 5

$$\frac{5}{10} = \frac{}{}$$

÷ 5

6

÷ ____

$$\frac{12}{15} = \frac{4}{}$$

÷ ____

7

÷ ____

$$\frac{16}{18} = \frac{}{9}$$

÷ ____

8

÷ ____

$$\frac{10}{16} = \frac{}{}$$

÷ ____

9

÷ ____

$$\frac{9}{12} = \frac{}{}$$

÷ ____

10

$$\frac{20}{25} = \frac{}{}$$

11

$$\frac{21}{27} = \frac{}{}$$

12 $\frac{10}{20} = \frac{5}{} = \frac{}{2} = \frac{}{16} = \frac{4}{} = \frac{3}{} = \frac{}{12}$

 ISBN: 9780170447331

Comparing fractions

- Comparing and ordering fractions is easiest if they all have the same denominator.

Example: Which is larger, $\frac{2}{3}$ or $\frac{3}{4}$?

One method is to compare pictures.

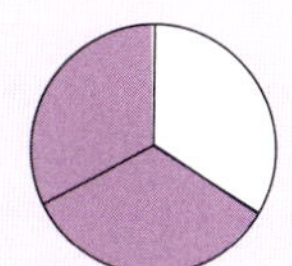

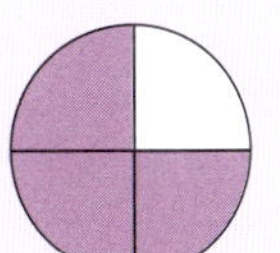

$\frac{3}{4}$ **looks** larger.

So, $\frac{3}{4}$ is larger than $\frac{2}{3}$.

Shade the diagrams and write which fraction is larger and which is smaller. The first one has been done for you.

1 $\frac{2}{6}$ $\frac{1}{4}$

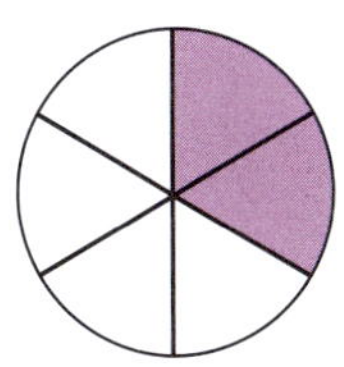

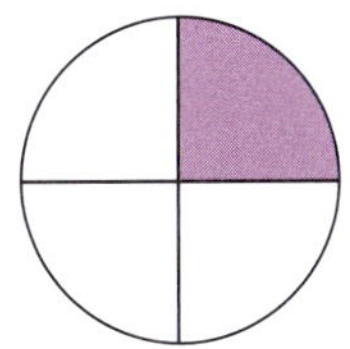

Larger ______ Smaller ______

2 $\frac{5}{7}$ $\frac{3}{5}$

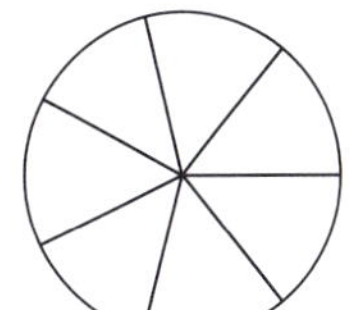

______ ______

3 $\frac{2}{5}$ $\frac{1}{3}$

______ ______

4 $\frac{4}{9}$ $\frac{2}{4}$

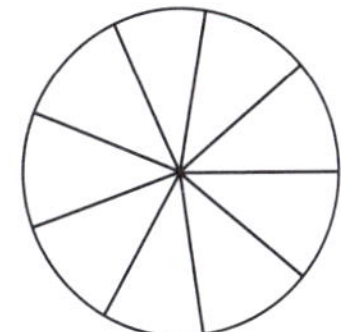

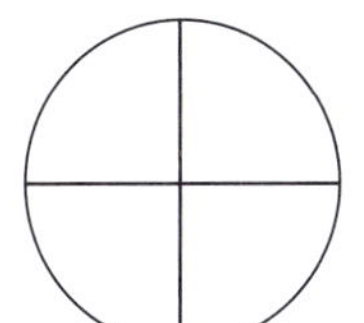

______ ______

ISBN: 9780170447331

5 $\frac{1}{4}$ $\frac{2}{7}$

6 $\frac{5}{8}$ $\frac{3}{5}$

7 $\frac{4}{10}$ $\frac{1}{3}$

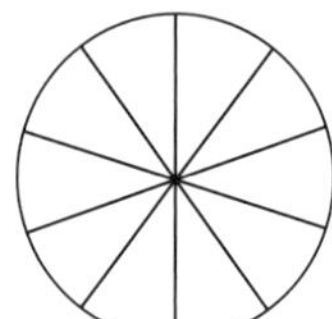

8 $\frac{3}{4}$ $\frac{4}{6}$

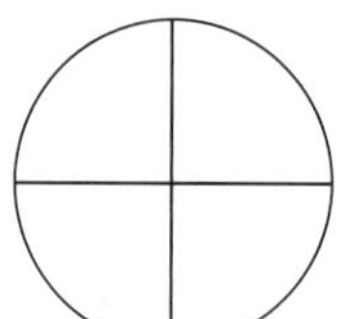

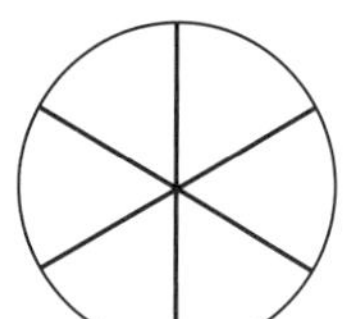

9 Shade these fractions and write them in ascending order (smallest to largest).

$\frac{4}{9}$

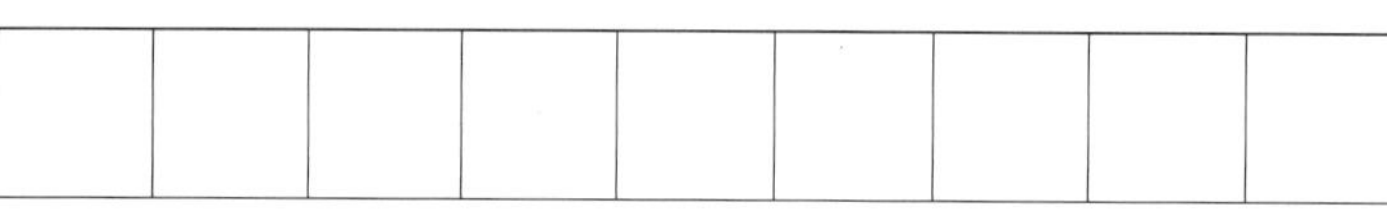

$\frac{1}{3}$

$\frac{2}{5}$

Smallest ________ ________ ________ Largest

 ISBN: 9780170447331

Converting between improper and mixed fractions

- An **improper** fraction is when the numerator is bigger than the denominator.

Example:

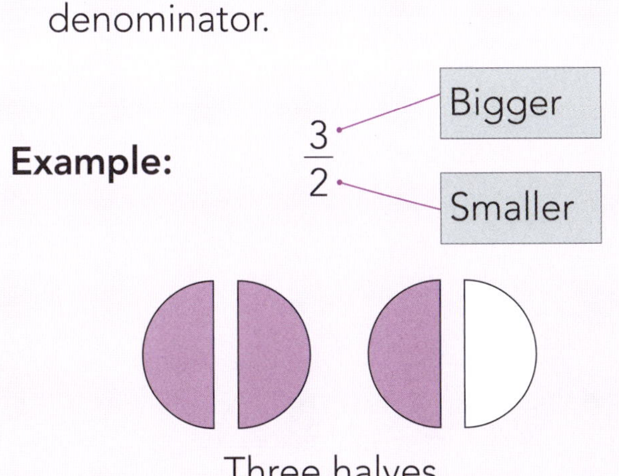

Three halves

- A **mixed** fraction is a combination of a whole number and a fraction.

Example:

$1\frac{1}{2}$

Whole number

Fraction

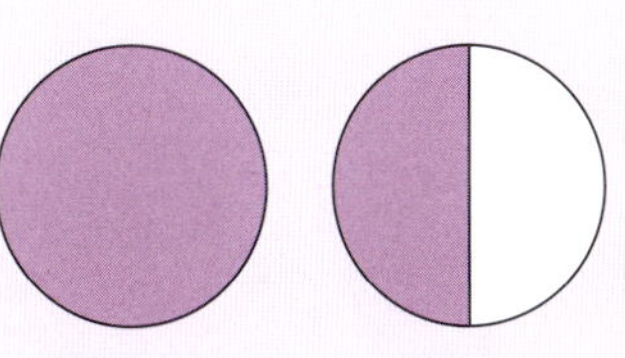

One and one half

Complete the table. The first row has been done for you.

	Improper fraction	Mixed fraction
1	$\frac{5}{4}$ Five quarters	$1\frac{1}{4}$ One and one quarter
2	___ ________________	___ ________________
3	___ ________________	___ ________________

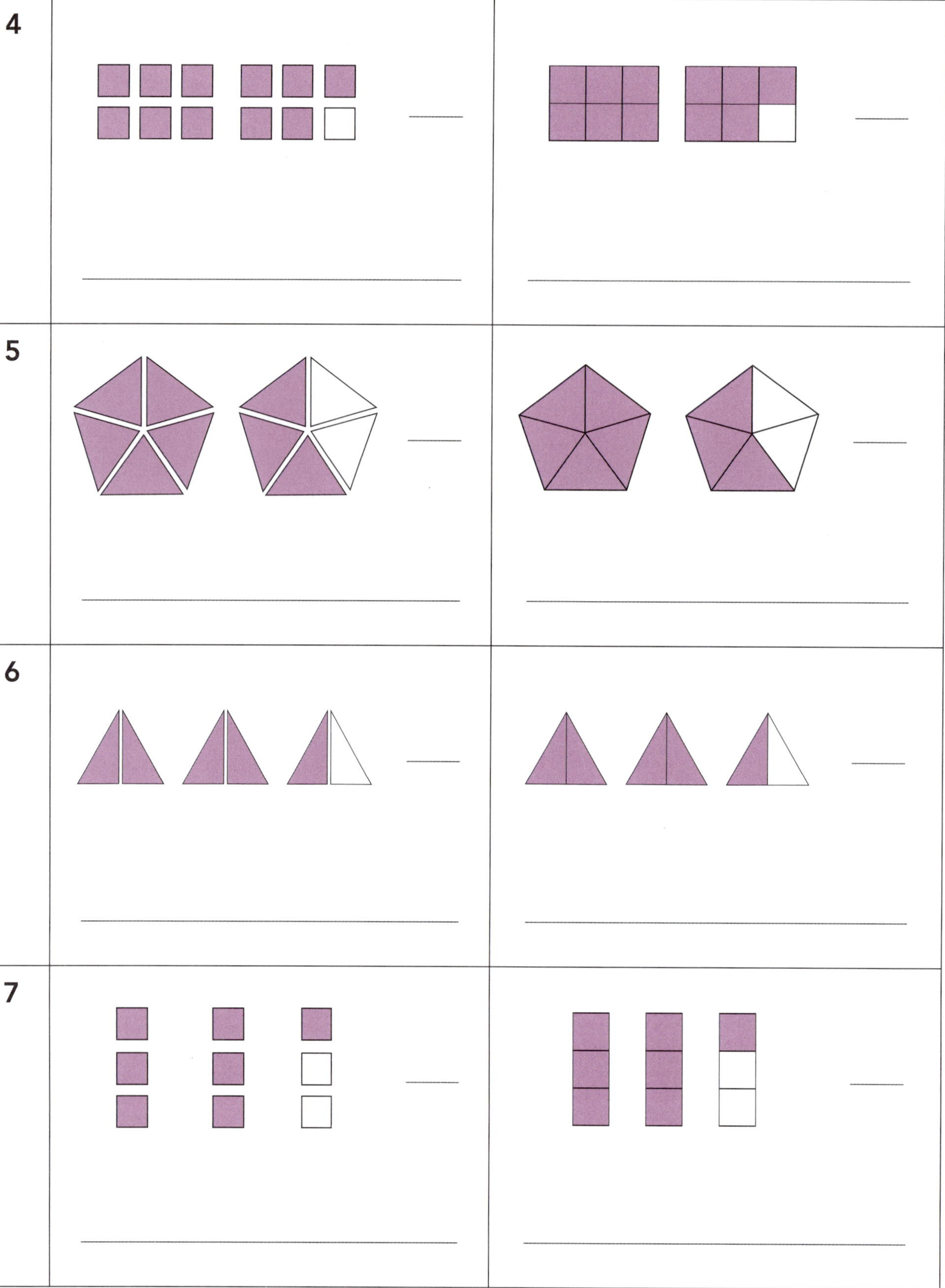

 ISBN: 9780170447331

Converting improper fractions to mixed fractions

Example: Convert $\frac{5}{3}$ into a mixed fraction.

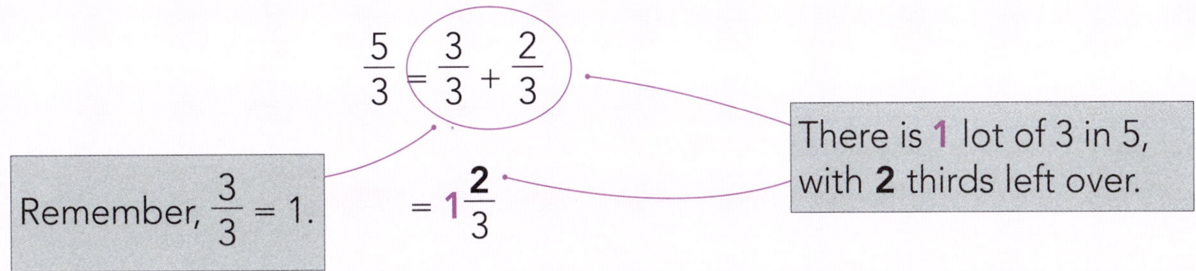

Complete the following.

8 $\frac{7}{4} = \frac{4}{4} + \frac{3}{4} = 1\frac{\square}{4}$

9 $\frac{9}{5} = \frac{5}{5} + \frac{4}{5} = 1\frac{\square}{5}$

10 $\frac{7}{6} = \frac{6}{6} + \frac{1}{6} = 1\frac{\square}{6}$

11 $\frac{5}{2} = \frac{2}{2} + \frac{2}{2} + \frac{1}{2} = 2\frac{\square}{2}$

12 $\frac{8}{3} = \frac{3}{3} + \frac{3}{3} + \frac{\square}{3} = 2\frac{\square}{3}$

13 $\frac{10}{7} = \frac{7}{7} + \frac{\square}{7} = 1\frac{\square}{7}$

14 $\frac{7}{2} = \frac{\square}{2} + \frac{\square}{2} + \frac{\square}{2} + \frac{\square}{2} = \square\frac{\square}{2}$

15 $\frac{9}{4} = \frac{\square}{4} + \frac{\square}{4} + \frac{\square}{4} = \square\frac{\square}{4}$

16 $\frac{10}{6} = \frac{\square}{6} + \frac{\square}{6} = \square\frac{\square}{6}$

17 $\frac{11}{3} = \frac{\square}{3} + \frac{\square}{3} + \frac{\square}{3} + \frac{\square}{3} = \square\frac{\square}{3}$

18 $\frac{6}{5} =$

19 $\frac{13}{7} =$

ISBN: 9780170447331

Converting mixed fractions to improper fractions

Example: Convert $1\frac{1}{3}$ into an improper fraction.

$$1\frac{1}{3} = \frac{3}{3} + \frac{1}{3}$$

There is **1** lot of **3** thirds, with **1** third left over.

$$= \frac{4}{3}$$

Complete the following.

20 $1\frac{3}{4} = \frac{4}{4} + \frac{3}{4} = \frac{\square}{4}$

21 $2\frac{1}{5} = \frac{5}{5} + \frac{5}{5} + \frac{1}{5} = \frac{\square}{5}$

22 $1\frac{2}{6} = \frac{6}{6} + \frac{2}{6} = \frac{\square}{6}$

23 $2\frac{2}{3} = \frac{3}{3} + \frac{3}{3} + \frac{2}{3} = \frac{\square}{3}$

24 $3\frac{1}{2} = \frac{2}{2} + \frac{2}{2} + \frac{2}{2} + \frac{1}{2} = \frac{\square}{2}$

25 $1\frac{3}{5} = \frac{\square}{5} + \frac{\square}{5} = \frac{\square}{5}$

26 $1\frac{1}{3} = \frac{\square}{3} + \frac{\square}{3} = \frac{\square}{3}$

27 $1\frac{5}{9} = \frac{\square}{9} + \frac{\square}{9} = \frac{\square}{9}$

28 $2\frac{1}{7} = \frac{\square}{7} + \frac{\square}{7} + \frac{\square}{7} = \frac{\square}{7}$

29 $2\frac{2}{4} = \frac{\square}{4} + \frac{\square}{4} + \frac{\square}{4} = \frac{\square}{4}$

30 $1\frac{1}{5} =$

31 $2\frac{3}{8} =$

ISBN: 9780170447331

Adding and subtracting fractions

With the same denominators

- To add or subtract fractions with the **same denominator**, you add or subtract the numerator. The denominator does not change.

Example: $\frac{1}{4} + \frac{1}{4} = \frac{2}{4}$

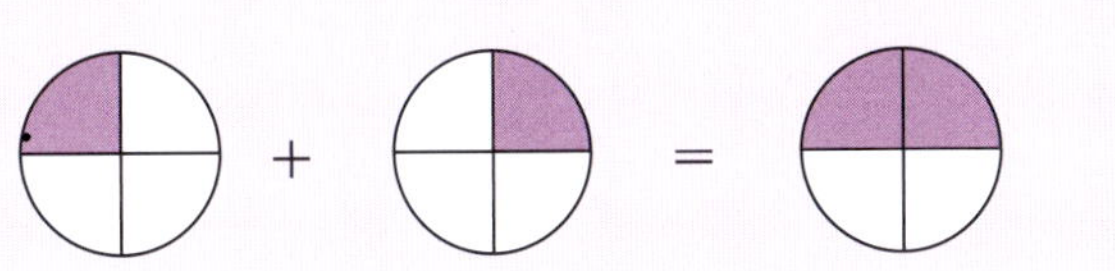

1 + 1 is 2.

The denominator remains a 4.

Complete the shading and fill in the missing numbers to add and/or subtract fractions.

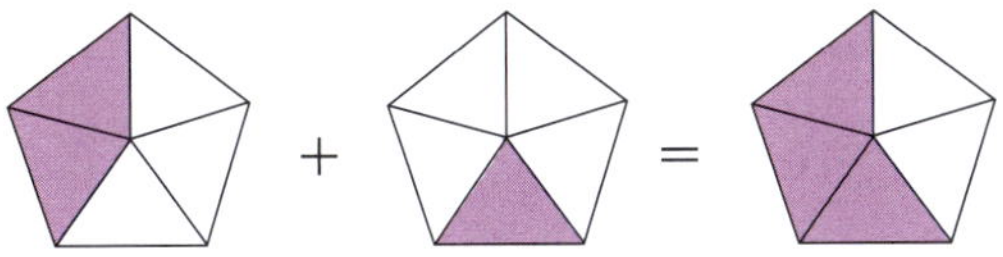

1 $\frac{2}{4} + \frac{}{4} = \frac{}{}$

2 $\frac{2}{5} + \frac{}{5} = \frac{}{5}$

3 ___ + ___ = ___

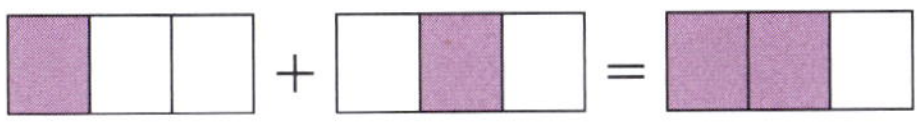

4 ___ − ___ = ___

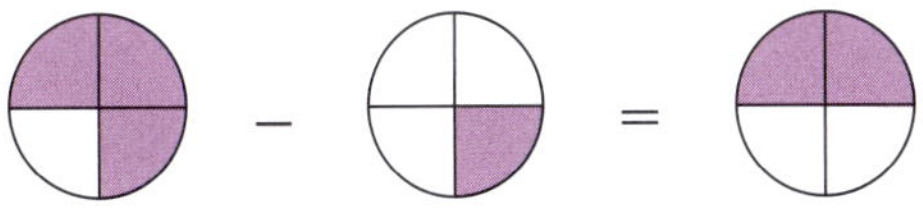

5 ___ + ___ = ___

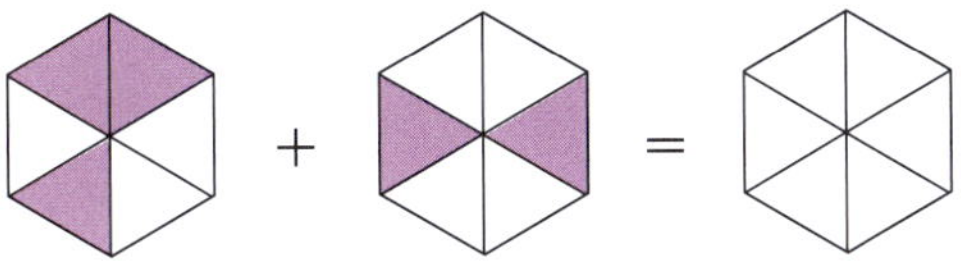

6 ___ − ___ = ___

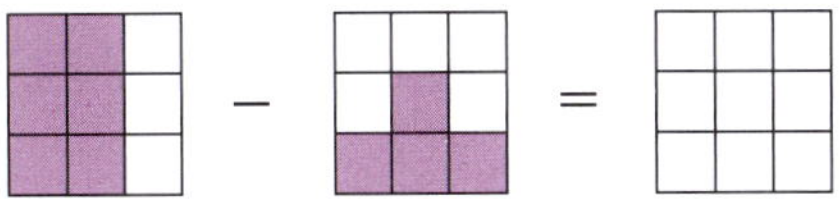

7 $\frac{3}{5} + \frac{1}{5} =$

8 $\frac{5}{8} - \frac{3}{8} =$

ISBN: 9780170447331

Multiplying fractions

- To multiply fractions, **multiply the numerators** and then **multiply the denominators**.
- Mixed fractions must be converted to improper fractions first.
- If possible, simplify the answer and write it as a mixed fraction.

Examples:

1 $\frac{1}{4} \times \frac{2}{3} = \frac{1 \times 2}{4 \times 3}$

Simplify. $= \frac{2}{12}$

$= \frac{1}{6}$

2 $\frac{5}{6} \times \frac{3}{2} = \frac{5 \times 3}{6 \times 2}$

$= \frac{15}{12} = \frac{5}{4}$

Convert to a mixed fraction. $= 1\frac{1}{4}$

Multiply these fractions and simplify when possible.

1 $\frac{2}{3} \times \frac{1}{2} = \frac{2 \times \_\_}{3 \times \_\_}$

$= \frac{\_\_}{6}$

$= \frac{\_\_}{3}$

2 $\frac{2}{5} \times \frac{3}{4} = \frac{\_\_}{5 \times \_\_}$

$= \frac{\_\_}{20}$

$= \frac{\_\_}{10}$

3 $\frac{1}{6} \times \frac{2}{4} =$ ____________

= ____________

= ____________

4 $\frac{2}{3} \times \frac{3}{5} =$ ____________

= ____________

= ____________

5 $\frac{1}{5} \times \frac{1}{4} =$ ____________

= ____________

= ____________

6 $\frac{5}{6} \times \frac{2}{3} =$ ____________

= ____________

= ____________

7 $\frac{1}{3} \times \frac{6}{7} =$ ____________

= ____________

= ____________

8 $\frac{3}{7} \times \frac{2}{6} =$ ____________

= ____________

= ____________

ISBN: 9780170447331

Fractions of a quantity

Remember that '**of**' means you must '**multiply**'.

Example:

$$\frac{1}{4} \text{ of } 12 = \frac{1}{4} \times 12$$
$$= \frac{1}{4} \times \frac{12}{1}$$
$$= \frac{12}{4}$$
$$= 3$$

Replace the word '**of**' with a **x** sign.

Remember that '12' means 12 wholes, or $\frac{12}{1}$.

Answer the following questions.

1 $\frac{1}{3}$ of 18 = $\frac{}{3} \times \frac{}{1}$

= $\frac{}{3}$

= ______

2 $\frac{1}{2}$ of 42 = $\frac{}{2} \times \frac{}{1}$

= $\frac{}{2}$

= ______

3 $\frac{1}{5}$ of 25 = ______

= ______

= ______

4 $\frac{1}{6}$ of 42 = ______

= ______

= ______

5 $\frac{2}{3}$ of 90 = ______

= ______

= ______

6 $\frac{3}{4}$ of 24 = ______

= ______

= ______

7 $\frac{3}{5}$ of 30 = ______

= ______

= ______

8 $\frac{3}{2}$ of 16 = ______

= ______

= ______

ISBN: 9780170447331

Decimals

Place value

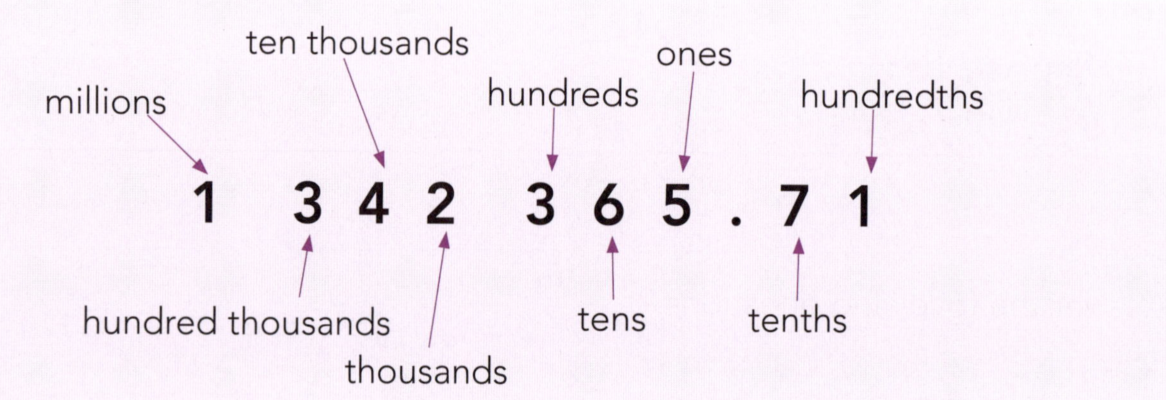

- When converting numerals to words, split the number where the **gaps** occur.

Examples: 1 Write 15 263 748 in words.

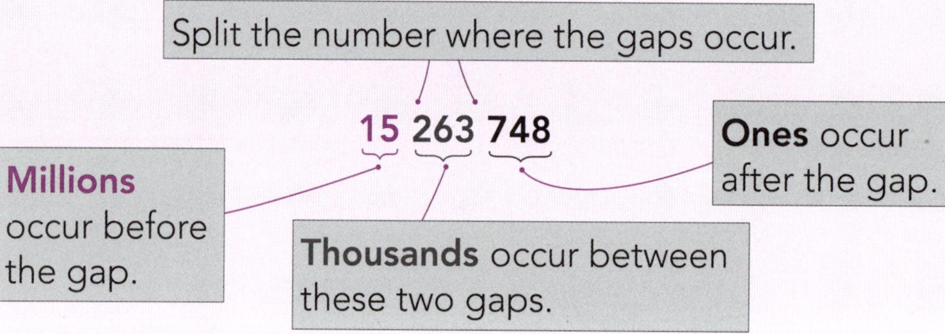

Fifteen million, **two hundred and sixty-three thousand**, **seven hundred and forty-eight** (ones).

2 Write 24 768.5 in words.

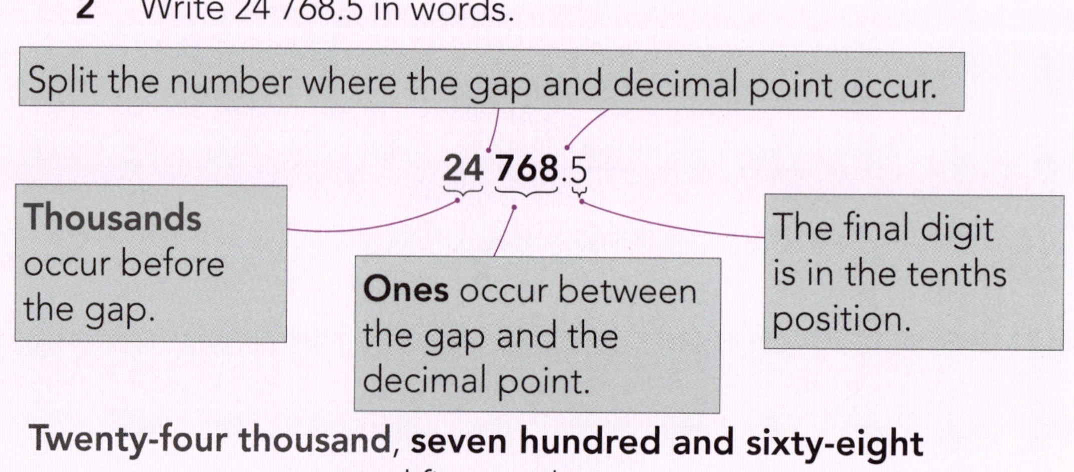

Twenty-four thousand, **seven hundred and sixty-eight** and five tenths.

ISBN: 9780170447331

Write down the value of the purple numerals as numbers and in words.

		Number	Words
1	4 619	4 000	
2	312 568		Ten thousand
3	7 056		
4	783 124		
5	381 266		
6	801 625		
7	8 924 119		
8	13 791		
9	8 320		

Write numbers using words.

10 2 381

____

11 90 162

____

12 500 487

____

13 103 200

____

14 670 054

____

15 502 007

____

16 1 600 151

____

ISBN: 9780170447331

Write numbers using numerals.

17 Four hundred thousand, three hundred and twenty-five __________

18 Eight thousand and sixty-eight 8 068

19 Thirty-four thousand, nine hundred and eleven __________

20 Two hundred and fifty-three thousand, four hundred and six __________

21 Two million, one hundred and thirty-seven thousand __________

22 Five hundred thousand and sixty-eight __________

23 Seventy thousand, four hundred and thirteen __________

The grids have 100 squares with a total area of 1. Shade each grid to match the decimal. Underneath each grid, write the decimal using words.

24 0.06

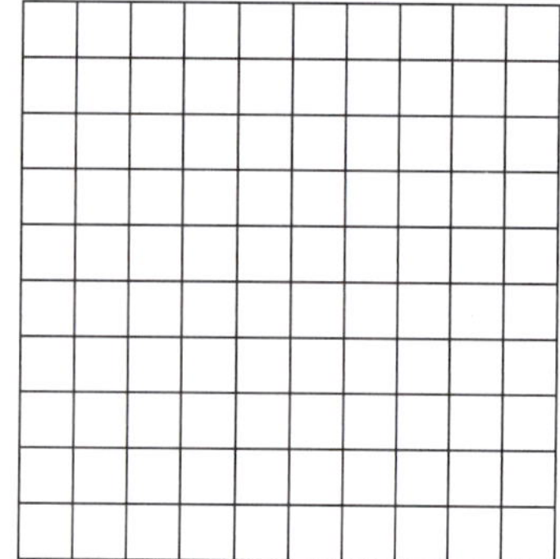

Six hundredths

25 0.2 (Remember 0.2 = 0.20)

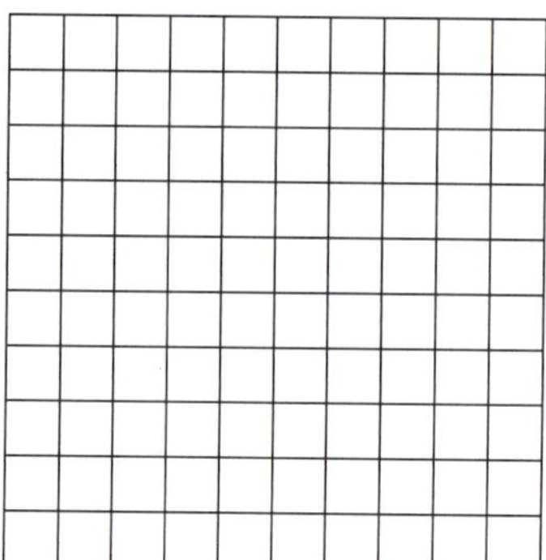

____

26 0.34

____

27 0.81

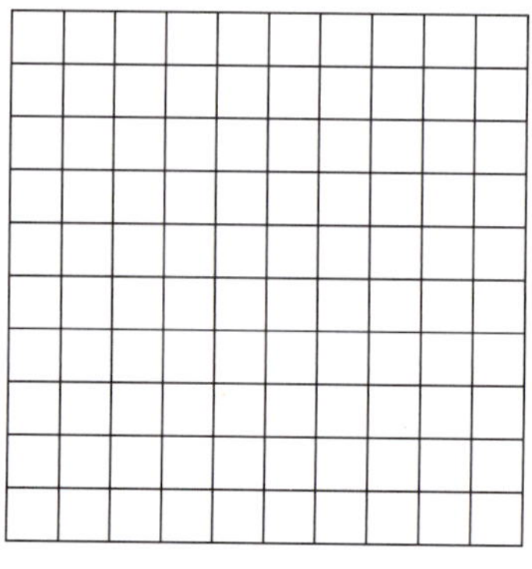

____

ISBN: 9780170447331

Write down the value of the purple numerals as decimals, fractions and in words.

		Decimal	Fraction		Words
28	0.91	0.9	$\frac{9}{10}$	$\frac{90}{100}$	Nine tenths or ninety hundredths
29	0.06		☒	____	
30	9.54	0.04	☒	$\frac{4}{100}$	
31	61.37		____	____	
32	0.05		☒	____	
33	8.20		____	____	
34	13.60		☒	____	
35	0.01		☒	____	

Write numbers using numerals.

36 Nine tenths ____________

37 Two hundredths 0.02

38 Sixteen hundredths ____________

39 Twelve tenths ____________

40 Eighty-three hundredths ____________

41 Five tenths and six hundredths ____________

42 One whole and nine hundredths ____________

43 Two wholes and thirty-four hundredths ____________

44 Four wholes and twenty-one hundredths ____________

45 Two wholes and seven hundredths ____________

ISBN: 9780170447331

Decimals on number lines

Here is how to work out the size of each gap between ticks on a number line.

Step 1: Calculate the **distance** between two **labelled** points.
Distance = 0.9 – 0.2 = **0.7**.

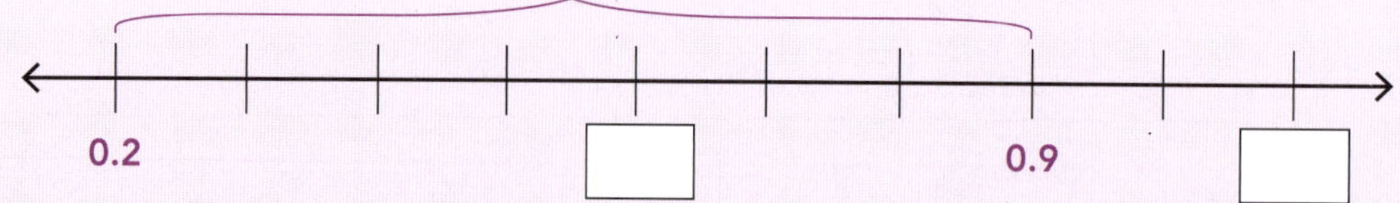

Step 2: Count the number of gaps between 0.2 and 0.9.
Number of gaps = 7.

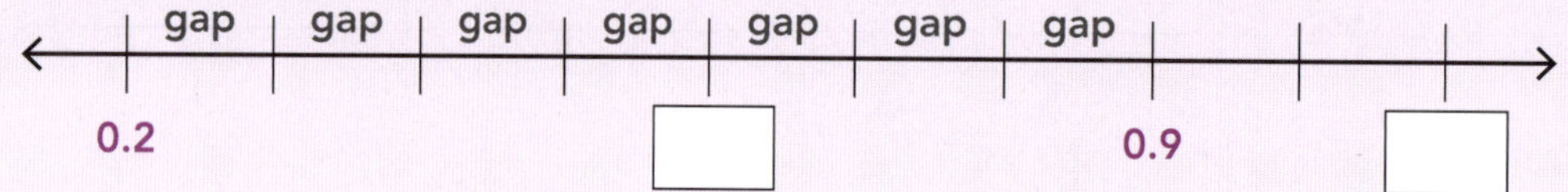

Step 3: Divide the distance by the number of gaps: $\frac{0.7}{7} = 0.1$.

Step 4: Add 0.1 after each gap along the number line.

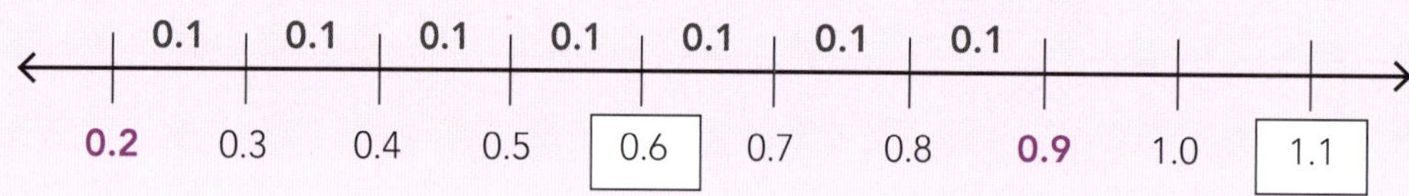

Calculate the size of each gap and write the missing decimals on the number lines.

1

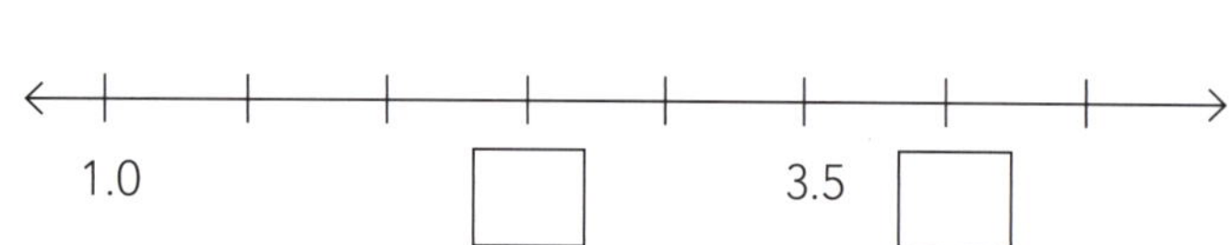

Size of gap = $\frac{\textbf{interval}}{\textbf{gaps}}$ =

$\frac{3.5 - 1.0}{5} = \frac{2.5}{5} = 0.5$

2

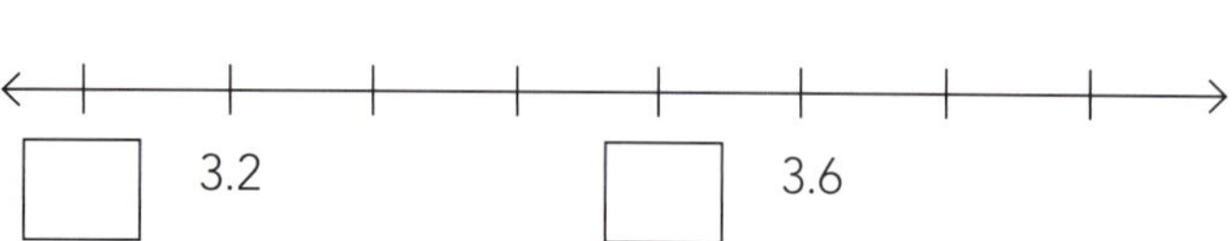

Size of gap =

____

____

 ISBN: 9780170447331

3

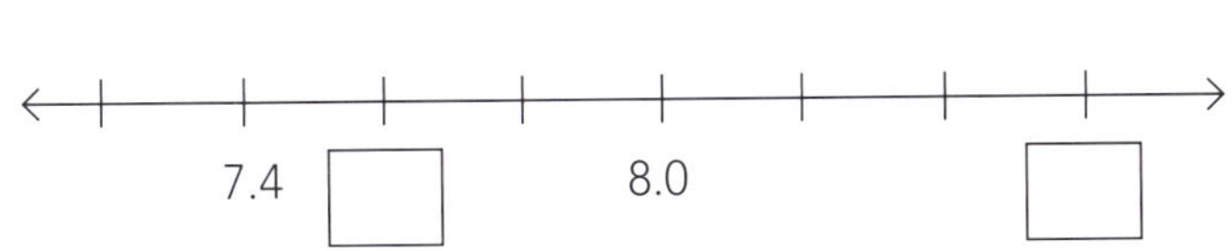

Size of gap =

4

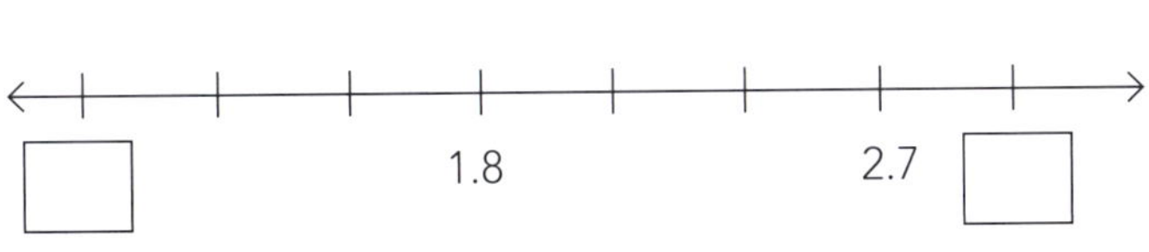

Size of gap =

5

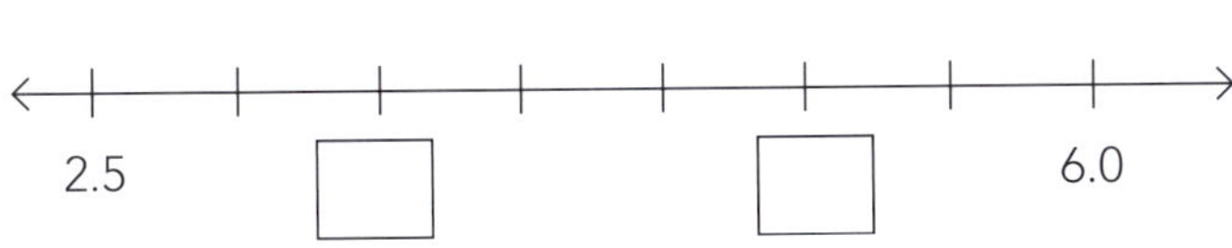

Size of gap =

6

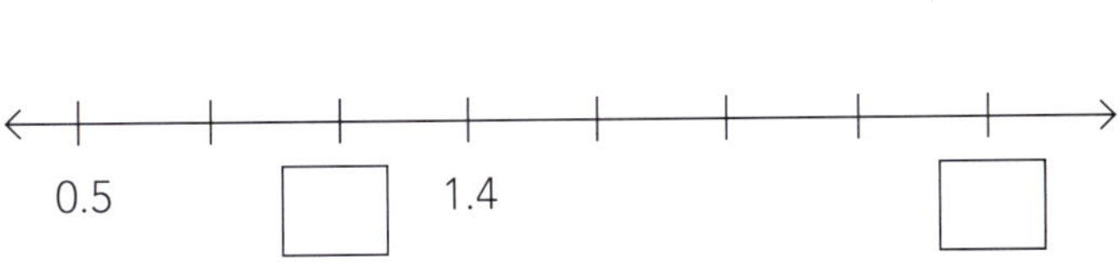

Size of gap =

7 From the table, select the most appropriate point for each position on the number line.

0.2	0.4	0.7
0.6	0.3	0.5
0.1	0.8	0.9

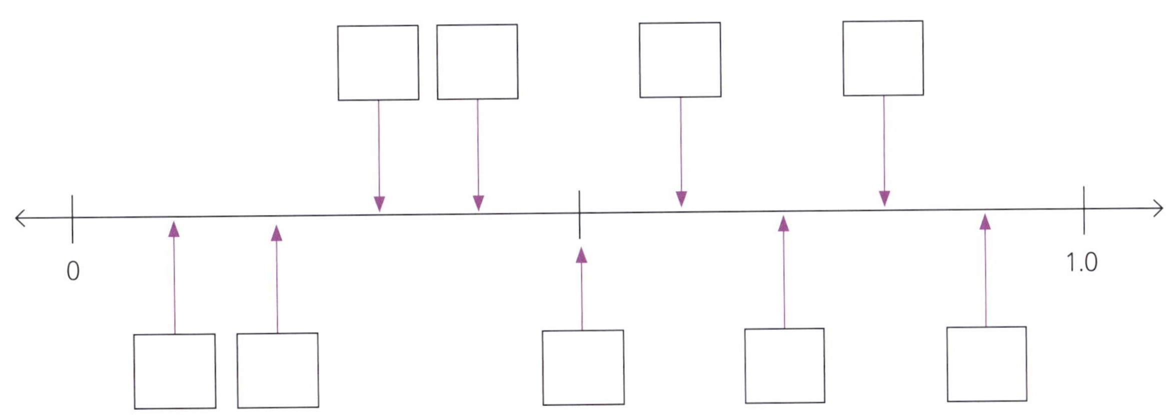

ISBN: 9780170447331

Comparing decimals

- To determine which decimal is largest, you need to consider the place values of each digit, starting from the left.

Examples:

1 Identify the larger number: **52.8** or **52.9**.

Step 1: Arrange the numbers with the decimal points lined up vertically:

52.8
52.9

The decimal points **must** be in line.

Step 2: Start at the left, and look for the **first pair of digits that is different**:

The first **two** digits are the same.

52|.**8**
52|.**9**

Step 3: Decide which is larger, **8** or **9**. 9 is larger, so 52.9 is the larger number.

2 Identify the larger number: **24.15** or **24.12**.

Step 1: Arrange the numbers with the decimal points lined up vertically:

24.15
24.12

Step 2: Start at the left, and look for the **first pair of digits that is different**:

The first **three** digits are the same.

24.1|**5**
24.1|**2**

Step 3: Decide which is larger, **5** or **2**. 5 is larger, so 24.15 is the larger number.

Highlight the larger number.

1 8.7
8.8

2 1 025
1 024

3 3.49
3.48

4 72.06
72.07

 ISBN: 9780170447331

5 123.0 or 123.1

6 11.11 or 11.10

7 1 001 or 1 010

8 0.09 or 0.10

9 11.0 or 10.1

10 5.01 or 5.10

11 100.1 or 101.0

12 0.21 or 0.12

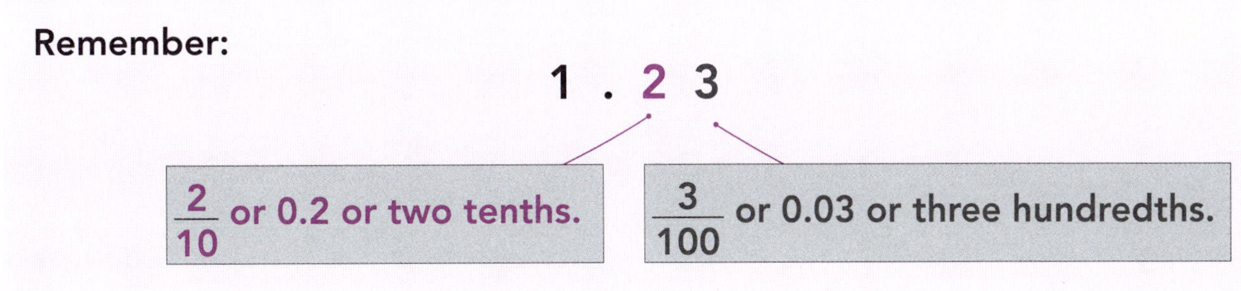

Write down the following decimals.

13 85.2 increased by three ones 88.2

14 21.7 increased by two tens ______

15 3 215 increased by two thousands ______

16 9.3 increased by two tenths ______

17 30.15 increased by four tenths ______

18 1 452.3 increased by two hundreds ______

19 654.15 increased by two hundredths ______

20 735.43 increased by five tenths ______

21 0.02 increased by three hundredths ______

22 70.4 increased by six tenths ______

ISBN: 9780170447331

Rearrange the following sets of digits in order to make the smallest and largest whole numbers possible. You must use all the digits.
Remember: zero can't be the first digit.

	Smallest	Largest
23 6 4 7	______	______
24 9 2 0 6	______	______
25 6 2 9 4 5	______	______
26 6 3 8 0	______	______
27 0 8 6 1 2 4	______	______

Place these decimals in ascending order (smallest to largest).

28 3.61 3.16 3.11 ______ ______ ______

29 0.11 0.01 0.10 ______ ______ ______

30 2.12 2.11 2.21 ______ ______ ______

31 0.87 0.88 0.78 ______ ______ ______

32 0.21 0.20 0.12 ______ ______ ______

Write the missing terms in the following number patterns. Then write in words the rule for each pattern.

33 0.07, 0.09, 0.11, ______, 0.15, ______

Rule: ______

34 3.25, 2.95, 2.65, ______, ______, ______

Rule: Subtract three tenths

35 10.81, 10.72, 10.63, ______, ______, ______

Rule: ______

ISBN: 9780170447331

Multiplying and dividing decimals by powers of 10

- To multiply a decimal by a power of 10, you move the decimal point to the **right**.

x 10^3 means the decimal place moves three places to the **right**.

Examples:

1 $1.8 \times 10^3 = 1\ .\ 8\ \ 0\ \ 0$
$= 1\ 800$

2 $0.009 \times 10^1 = 0\ .\ 0\ 0\ 9$
$= 0.09$

Remember, 1 800 is the same as 1 800**.** — you do not need to write the decimal point at the end.

Calculate these without using a calculator.

1 $12.5 \times 10^1 =$ ____________

2 $7.8 \times 10^2 =$ ____________

3 $8.46 \times 10^3 =$ ____________

4 $0.63 \times 10^1 =$ ____________

5 $927.5 \times 10^3 =$ ____________

6 $0.006 \times 10^3 =$ ____________

- To divide a decimal by a power of 10, you move the decimal point to the **left**.

÷ 10^3 means the decimal place moves three places to the **left**.

Examples:

1 $2.3 \div 10^3 = 0\ \ 0\ \ 0\ \ 2\ .\ 3$
$= 0.0023$

2 $654 \div 10^1 = 6\ \ 5\ \ 4\ .$
$= 65.4$

Calculate these without using a calculator.

7 $53.2 \div 10^1 =$ ____________

8 $177.5 \div 10^2 =$ ____________

9 $5\ 800 \div 10^3 =$ ____________

10 $412.9 \div 10^1 =$ ____________

11 $0.6 \div 10^2 =$ ____________

12 $17\ 300 \div 10^4 =$ ____________

13 $0.15 \div 10^1 =$ ____________

14 $100\ 000 \div 10^2 =$ ____________

ISBN: 9780170447331

Using decimals to compare fractions

You can compare fractions by:
1 converting the fractions to decimals using your calculator,
2 then comparing by identifying which number is larger.

Examples: Identify which fraction is larger out of the two.

1 $\frac{3}{4}$ or $\frac{4}{5}$

Use a button that looks like

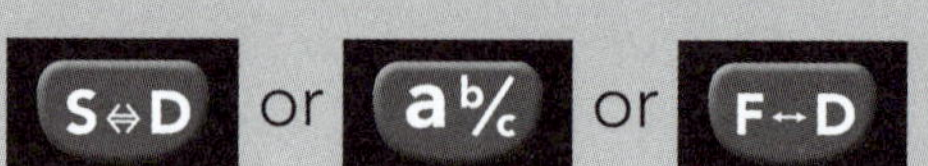

to turn the fraction into a decimal.

$\frac{3}{4} = 0.|75$

$\frac{4}{5} = 0.|8$

Write the decimals with their points lined up.

Draw a vertical line before the first numerals that are different. Compare these numerals.

8 is larger than **7**, so $\frac{4}{5}$ is larger than $\frac{3}{4}$.

2 $\frac{3}{5}$ or $\frac{5}{8}$

$\frac{3}{5} = 0.6|0$

$\frac{5}{8} = 0.6|25$

Sometimes you will need to add a 0 so you have numerals to compare.

2 is larger than **0**, so $\frac{5}{8}$ is larger than $\frac{3}{5}$.

Convert these pairs of fractions to decimals using your calculator, and then decide which is larger.

1 $\frac{1}{4} =$ ______ $\frac{13}{50}$ is larger than ______

$\frac{13}{50} =$ 0.26

2 $\frac{3}{20} =$ ______ ______ is larger than ______

$\frac{1}{5} =$ ______

 ISBN: 9780170447331

3 $\frac{3}{4}$ = ________

$\frac{19}{25}$ = ________

______ is larger than ______

4 $\frac{17}{20}$ = ________

$\frac{4}{5}$ = ________

______ is larger than ______

5 $\frac{11}{20}$ = ________

$\frac{13}{25}$ = ________

______ is larger than ______

6 $\frac{7}{25}$ = ________

$\frac{1}{4}$ = ________

______ is larger than ______

7 $\frac{3}{4}$ = ________

$\frac{39}{50}$ = ________

______ is larger than ______

8 $\frac{7}{8}$ = ________

$\frac{43}{50}$ = ________

______ is larger than ______

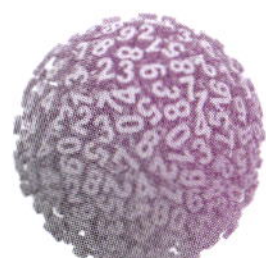

Challenge 2

Convert the fractions to decimals using your calculator, then place the fractions in ascending order (smallest to largest).

1 $\frac{9}{20}$ = ________

$\frac{2}{5}$ = ________

$\frac{11}{25}$ = ________

________ ________ ________

Smallest Largest

2 $\frac{16}{125}$ = ________

$\frac{31}{250}$ = ________

$\frac{1}{8}$ = ________

________ ________ ________

Smallest Largest

ISBN: 9780170447331

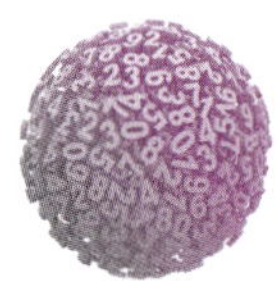

Percentages

- Percentages are a way of expressing a number **out of 100**.

Example:

32 out of 100 squares are purple.

$\frac{32}{100}$ is the same as 32%.

68 out of 100 squares are white.

This means 68% are white.

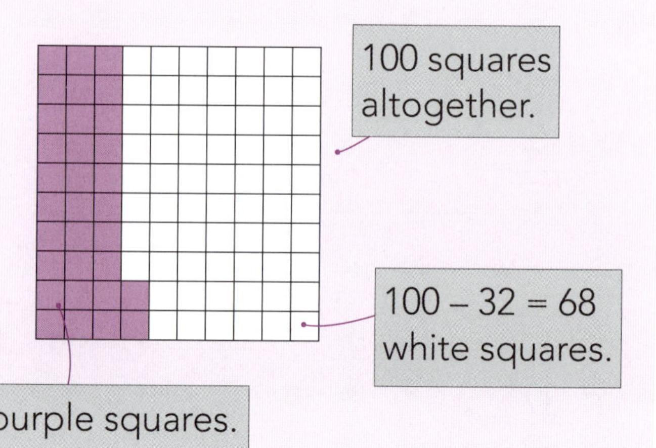

Fill in the gaps in the following.

1

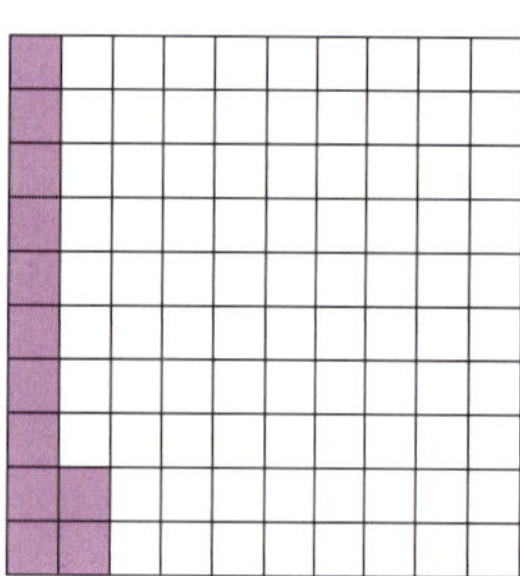

Percentage **purple** ____________

Percentage white ____________

2

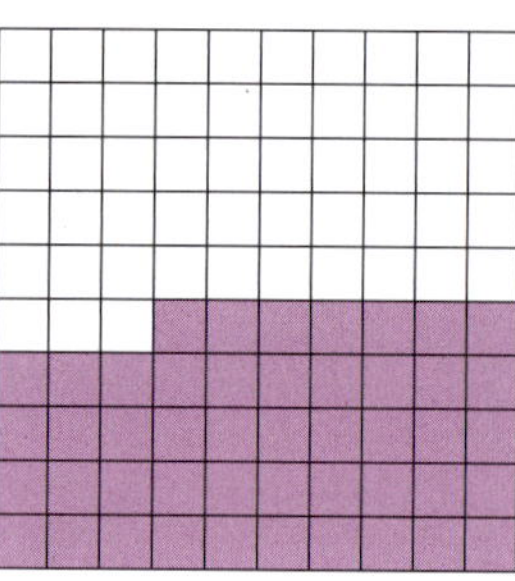

Percentage **purple** ____________

Percentage white ____________

3

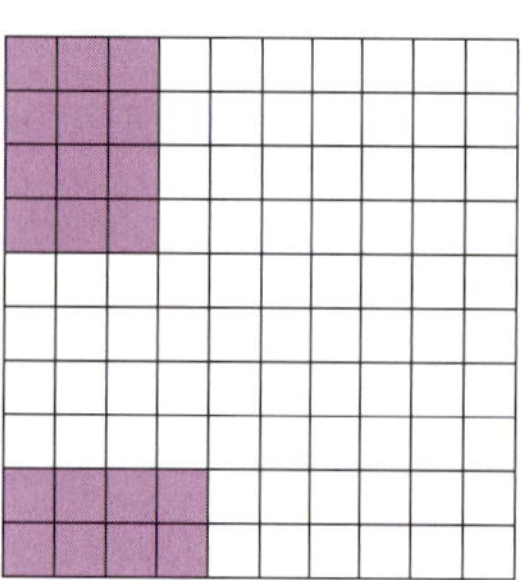

Percentage **purple** ____________

Percentage white ____________

4

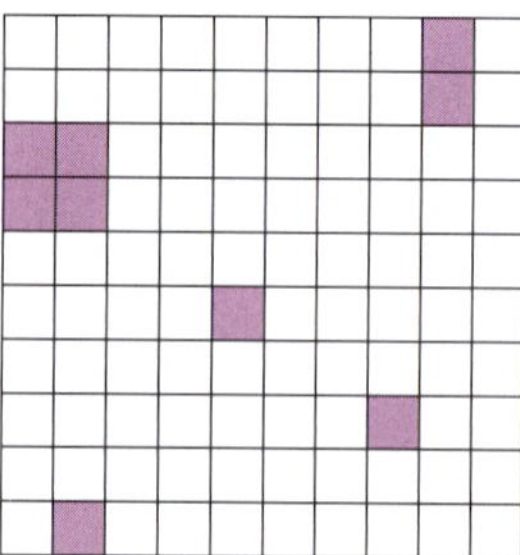

Percentage **purple** ____________

Percentage white ____________

ISBN: 9780170447331

5

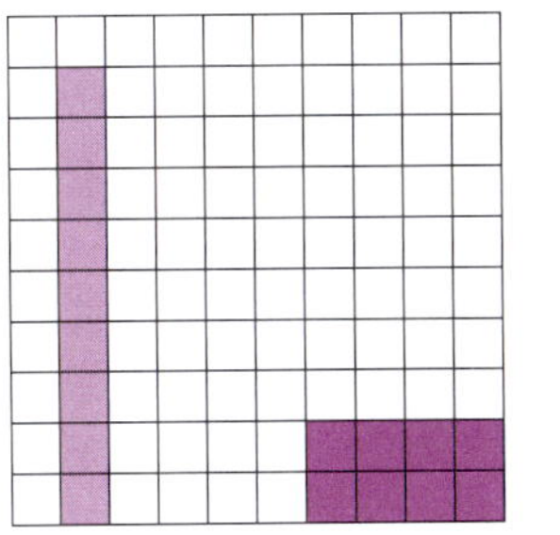

Percentage dark purple ______

Percentage light purple ______

Percentage white ______

Total: ______

6

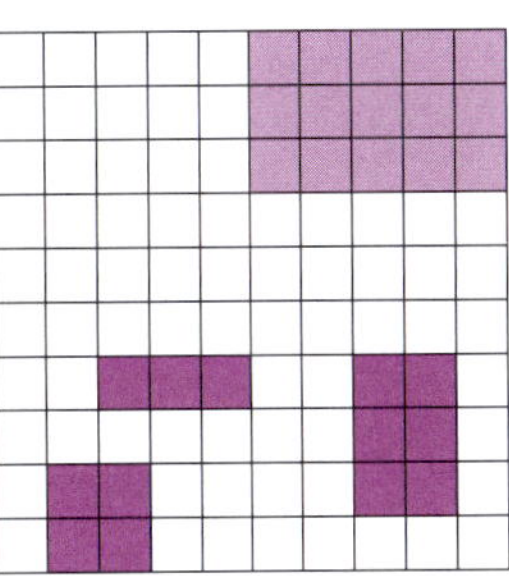

Percentage dark purple ______

Percentage light purple ______

Percentage white ______

Total: ______

7

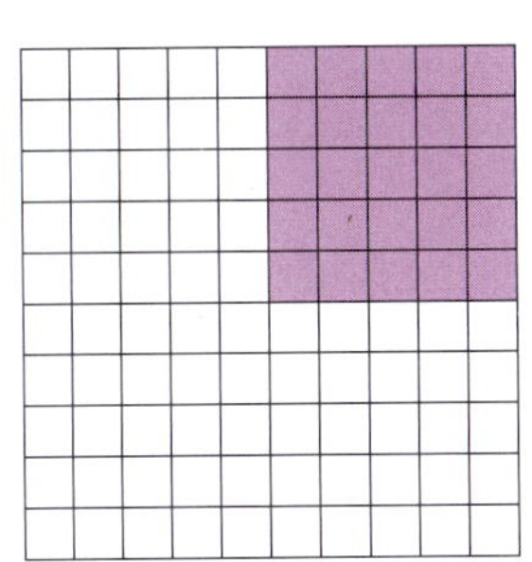

Percentage light purple ______

Fraction light purple $\frac{}{100}$

Fraction light purple $\frac{}{4}$

8

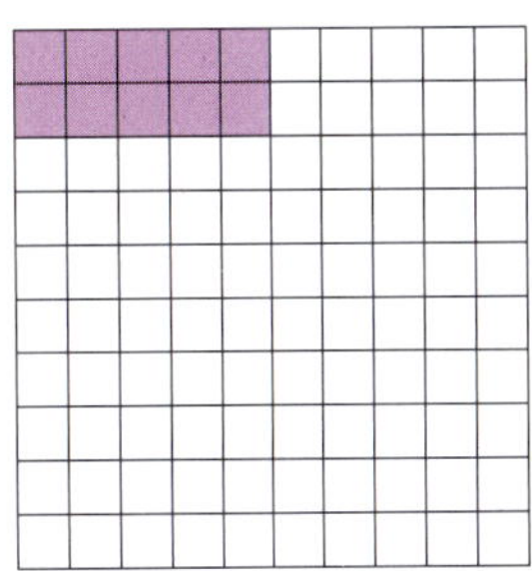

Percentage light purple ______

Fraction light purple $\frac{}{100}$

Fraction light purple $\frac{}{10}$

9

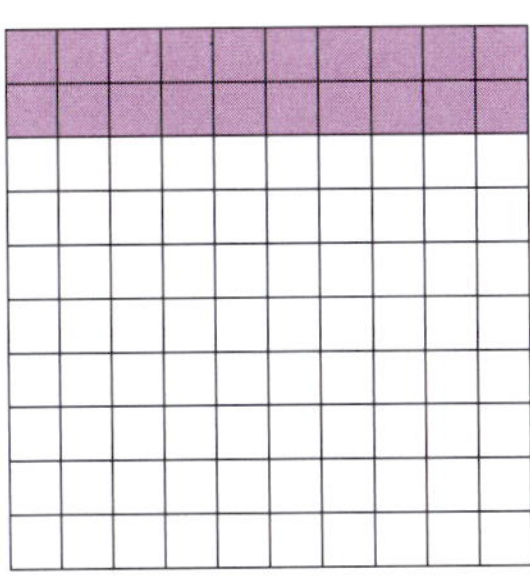

Percentage light purple ______

Fraction light purple $\frac{}{100}$

Fraction light purple $\frac{}{5}$

10

Percentage light purple ______

Fraction light purple $\frac{}{100}$

Fraction light purple $\frac{}{5}$

ISBN: 9780170447331

Changing fractions to percentages

To change a fraction to a percentage: multiply by 100.

Example:

24 out of 50 squares are purple.

$\frac{24}{50}$ needs to be multiplied by 100 to make a percentage.

$\frac{24}{50}$ **x 100** = 48%, so 48% of the squares are purple.

$\frac{26}{50}$ of the squares are white.

$\frac{26}{50}$ **x 100** = 52%, so 52% of the squares are white.

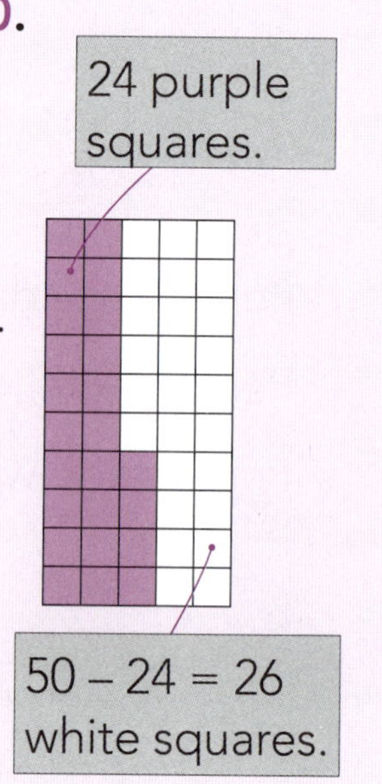

State the percentages shown in these diagrams.

1

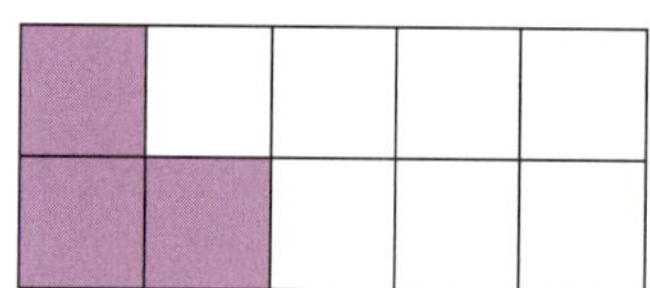

Percentage purple

$\frac{}{10}$ x **100** = ______

Percentage white

$\frac{}{10}$ x **100** = ______

2

Percentage purple

$\frac{}{20}$ x **100** = ______

Percentage white

$\frac{}{20}$ x **100** = ______

3

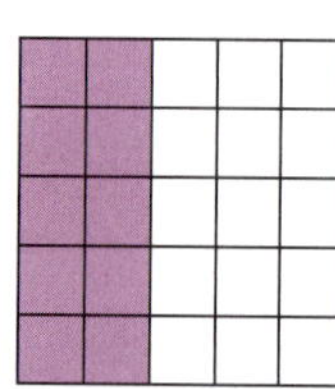

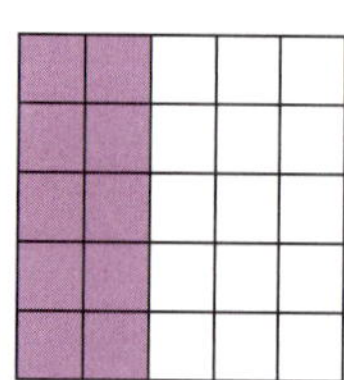

Percentage purple

$\frac{}{25}$ x **100** = ______

Percentage white

$\frac{}{25}$ x **100** = ______

4

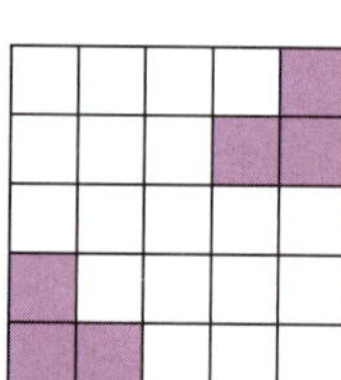

Percentage purple

—— x **100** = ______

Percentage white

—— x **100** = ______

 ISBN: 9780170447331

5

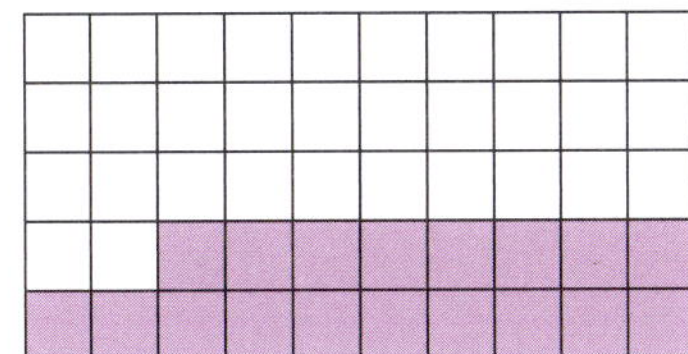

Percentage purple

___ x ______ = ______

Percentage white

___ x ______ = ______

6

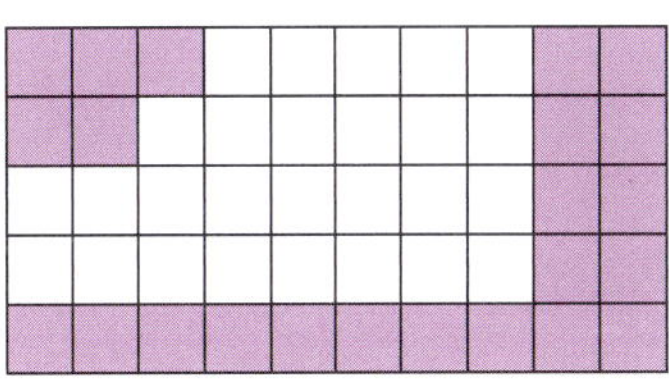

Percentage purple

___ x ______ = ______

Percentage white

___ x ______ = ______

7

Percentage purple

___ x ______ = ______

Percentage white

___ x ______ = ______

8

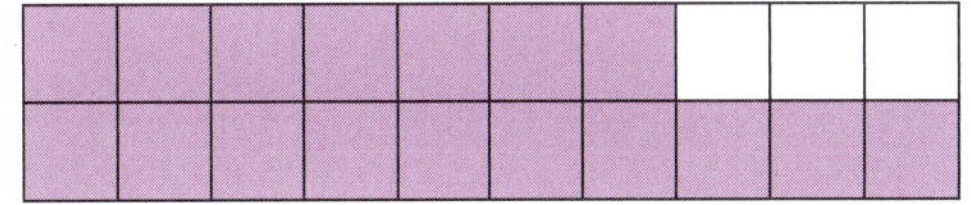

Percentage purple

___ x ______ = ______

Percentage white

___ x ______ = ______

Convert these fractions into percentages.

9 $\frac{9}{20}$: $\frac{9}{20}$ x 100 = ________ %

10 $\frac{1}{4}$: — x 100 = ________ %

11 $\frac{7}{10}$: ____________________

12 $\frac{3}{5}$: ____________________

13 $\frac{21}{50}$: ____________________

14 $\frac{13}{25}$: ____________________

15 $\frac{38}{40}$: ____________________

16 $\frac{17}{20}$: ____________________

17 $\frac{3}{4}$: ____________________

18 $\frac{18}{60}$: ____________________

Changing percentages to fractions

To change a percentage to a fraction: divide by 100, then simplify the fraction.

Don't forget to simplify the fraction.

Examples: 1 $60\% = \frac{60}{100} = \frac{3}{5}$ (÷ 100)

2 $34\% = \frac{34}{100} = \frac{17}{50}$ (÷ 100)

Convert these percentages into simplified fractions.

1 $20\% = \frac{\quad}{100} = \frac{\quad}{5}$ (÷ 100)

2 $75\% = \frac{\quad}{100} = \frac{\quad}{4}$ (÷ 100)

3 $17\% = \frac{\quad}{100}$ (÷ 100)

4 $35\% = \frac{\quad}{100} = \frac{\quad}{20}$ (÷ 100)

5 $90\% = \frac{\quad}{100} = \frac{9}{\quad}$

6 $24\% = \frac{\quad}{100} = \frac{6}{\quad}$

7 $62\% = \frac{\quad}{100} = \frac{31}{\quad}$

8 $4\% = \frac{\quad}{100} = \frac{1}{\quad}$

9 7% = ____________________

10 8% = ____________________

11 48% = ____________________

12 85% = ____________________

13 1% = ____________________

14 100% = ____________________

 ISBN: 9780170447331

Changing decimals to percentages

To change a decimal to a percentage: multiply by 100.

Examples: 1 0.30 = 30% (× 100) **2** 0.75 = 75% (× 100)

Notice that **multiplying** by 100 is the same as moving the decimal point two places to the **right**.

0 . 3 0 = 30%

Convert the following decimals into percentages.

1 0.15 = ______ (× 100) **2** 0.90 = ______ (× 100)

3 0.24 = ______ **4** 0.82 = ______

5 0.17 = ______ **6** 0.99 = ______

7 0.43 = ______ **8** 0.16 = ______

9 0.50 = ______ **10** 0.35 = ______

11 0.10 = ______ **12** 0.09 = ______

13 0.04 = ______ **14** 0.01 = ______

ISBN: 9780170447331

Changing percentages to decimals

To change a percentage to a decimal: divide by 100.

Notice that **dividing** by 100 is the same as moving the decimal point two places to the **left**.

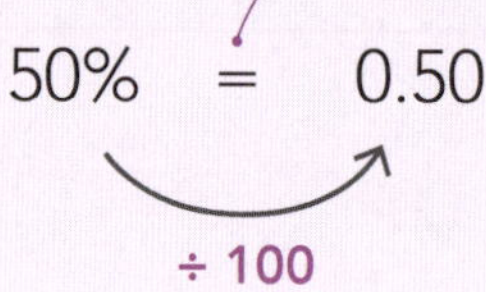

Examples: 1 50% = 0.50 (÷ 100) **2** 18% = 0.18 (÷ 100)

Convert the following percentages into decimals.

1 25% = ______ (÷ 100)

2 60% = ______ (÷ 100)

3 90% = ______

4 17% = ______

5 85% = ______

6 24% = ______

7 32% = ______

8 63% = ______

9 10% = ______

10 72% = ______

11 5% = ______

12 99% = ______

13 100% = ______

14 1% = ______

ISBN: 9780170447331

Converting between fractions, decimals and percentages

Fill in the gaps.

	Fraction	Decimal	Percentage
1	$\frac{1}{2}$		
2			30%
3		0.4	
4	$\frac{1}{4}$		
5		0.95	
6			80%
7	$\frac{1}{5}$		
8		0.75	
9			10%
10	$\frac{10}{10}$		
11		0.99	
12	$\frac{3}{20}$		
13			16%
14		0.01	

Calculating percentages

- Remember, 'percent' means out of 100, so turning a fraction into a percentage means **multiplying by 100**.

Example: There are 35 students in the choir. Fourteen of them are boys. What percentage of the class are boys?

$$\% = \frac{14}{35} \times 100 = 40\%$$

Write **14** out of 35 as a fraction.

x by 100.

Write the following amounts as percentages.

1 6 out of 8 $= \frac{}{8} \times 100$

= ____________

2 10 out of 50 $= \frac{}{50} \times 100$

= ____________

3 19 out of 20 = ____________

= ____________

4 350 out of 1 000 = ____________

= ____________

5 24 out of 60 = ____________

= ____________

6 7 out of 700 = ____________

= ____________

7 12 out of 25 = ____________

= ____________

8 42 out of 150 = ____________

= ____________

9 A cucumber which weighs 200 g contains 192 g of water. What percentage of the cucumber is water?

____

10 An NBA basketball lasts for 10 000 bounces. The number of bounces in an average NBA game is 2 650. What percentage of the life of an NBA ball is used up during a single NBA basketball game?

____

ISBN: 9780170447331

Finding percentages of amounts

- There are many ways of doing this.

Remember, to convert you divide by 100.

Examples:

1 Find 25% of 60.

Either: convert the percentage to a decimal: 25% **of** 60 = 0.25 **x** 60
= 15

Remember, '**of**' means **x**.

Or: convert the percentage to a fraction: 25% **of** 60 = $\frac{25}{100}$ **x** 60
= 15

- It is also possible to use the % button on your **calculator**.
- Not all calculators are the same, so you will need to experiment until you find how yours works. Two possibilities are shown in the box below.

2 Find 30% of $90.

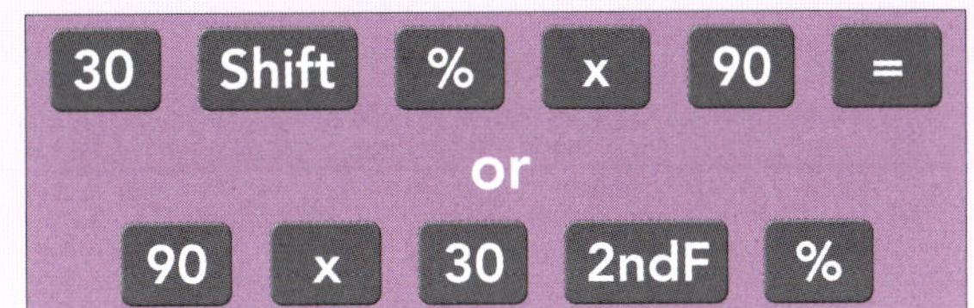

So, 30% of 90 = 27.

Calculate these.

1 10% of 70 = ____________
= ____________

2 40% of 45 = ____________
= ____________

3 25% of 600 = ____________
= ____________

4 80% of 650 = ____________
= ____________

5 15% of 80 = ____________
= ____________

6 75% of 24 = ____________
= ____________

ISBN: 9780170447331

7 4% of 1 200 = ______________

= ______________

8 25% of 40 000 = ______________

= ______________

9 1% of 750 = ______________

10 99% of 1 200 = ______________

11 55% of 78 = ______________

12 8% of 6 400 = ______________

13 **a** The sign on a shop window says '20% off everything'. How much will be taken off the price of a sweatshirt that normally costs $80?

____

b Calculate its sale price.

____

14 **a** Another shop advertises that all prices will be reduced by 25%. A phone normally costs $450. How much will be taken off the price of the phone?

____

b Calculate the reduced price of the phone.

____

15 Koalas sleep for 75% of the day because gum leaves take such a long time to be digested. For how many hours does a koala sleep during each day?

____

16 8% of people have an extra rib. If your school has 750 students, how many would be expected to have an extra rib?

____

17 A cow produces (farts) 120 kg of methane a year. A human produces 0.1% of the amount of methane that a cow produces. How much methane do you produce in a year?

____

 ISBN: 9780170447331

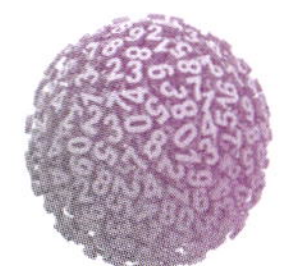

Rounding

- Often we need to round numbers to sensible and/or meaningful values.
- Never round until **after** you have completed your calculations.

Rounding to whole numbers

Locate the digit you have to round to. Is the digit to its **right** 5 or more?

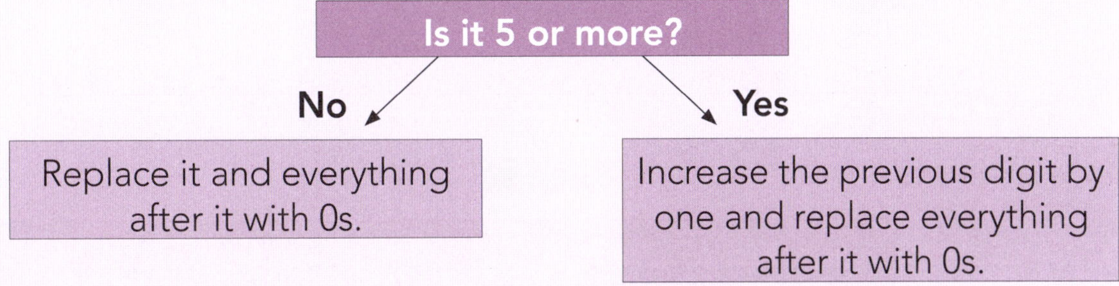

Examples: 1 Round 7**8**6 to the nearest **ten**.

Is 7**8**6 closer to 780 or 790?

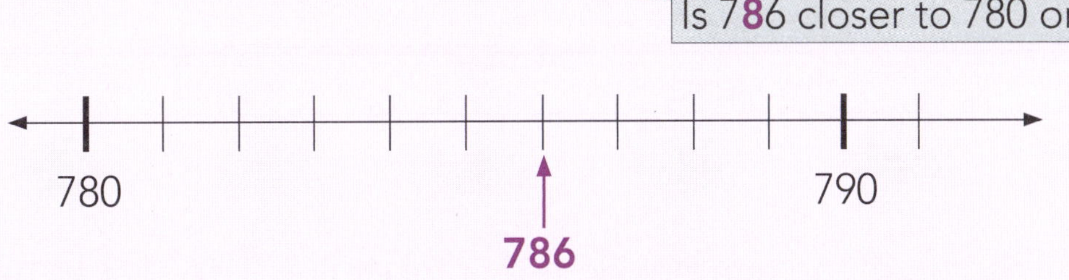

The digit to the right of the **tens** digit (**8**) in 7**8**6 is the **6**.
So 786 rounded to the nearest ten is 790.

2 Round 1 **5**49 to the nearest **hundred**.

Is 1 **5**49 closer to 1 500 or 1 600?

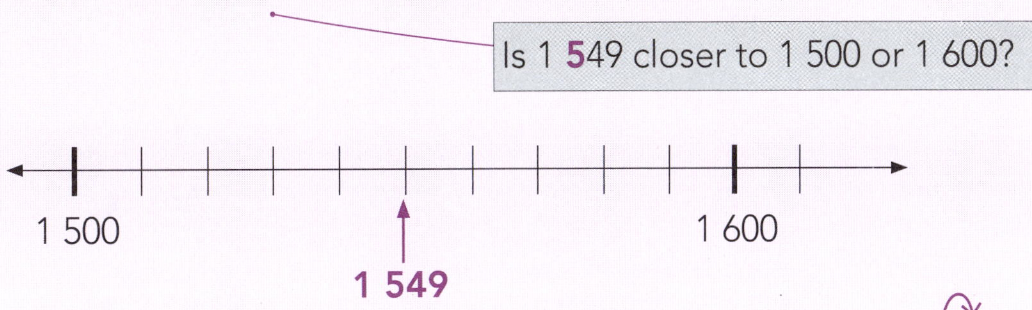

The digit to the right of the **hundreds** digit (**5**) in 1 **5**49 is the **4**.
So 1 549 rounded to the nearest hundred is 1 500.

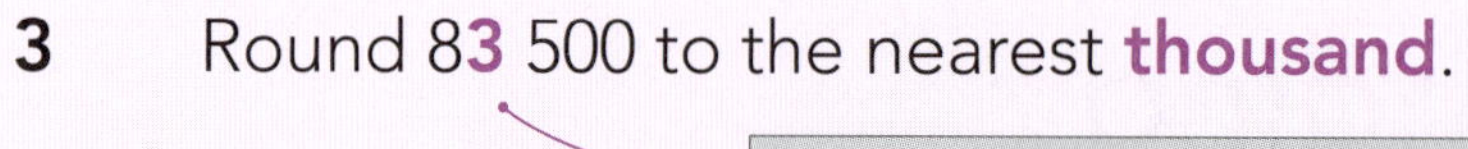

3 Round 83 500 to the nearest **thousand**.

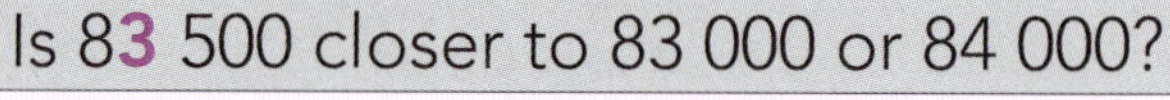

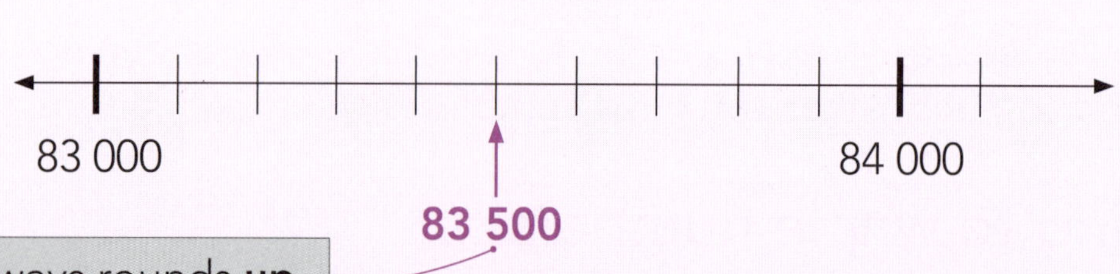

The digit to the right of the **thousands** digit (3) in 83 500 is the **5**.
So 83 500 rounded to the nearest thousand is 84 000.

More examples:

	Rounded to the nearest:	Last required digit	Answer
39	ten	39	40
1 523	hundred	1 523	1 500
120 973	thousand	120 973	121 000

Mark the position of the number on the number line. Use that to help you round these numbers to the nearest ten.

1 47 Answer ______________

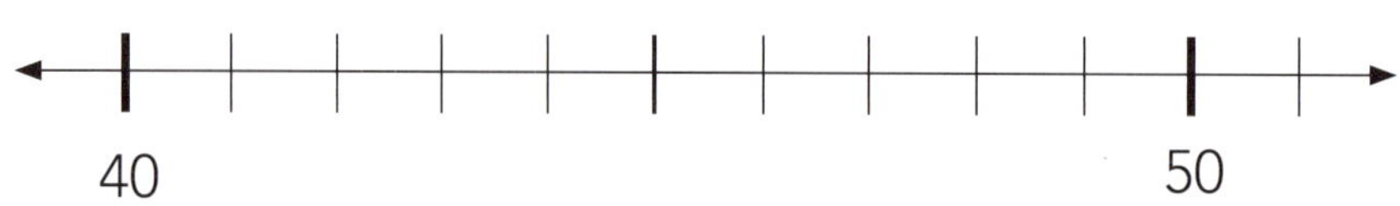

2 674 Answer ______________

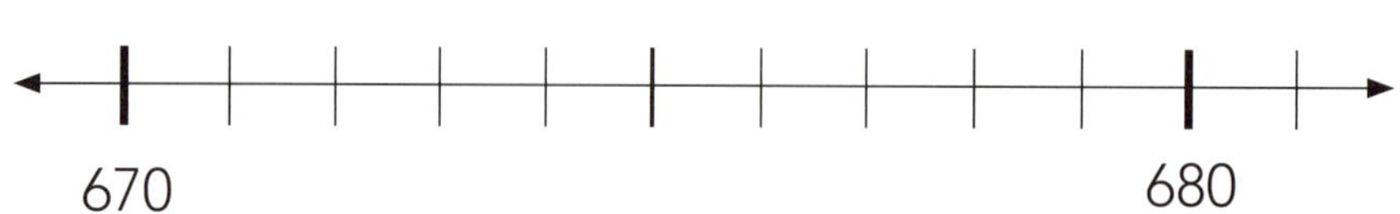

 ISBN: 9780170447331

3 1 385 Answer ____________

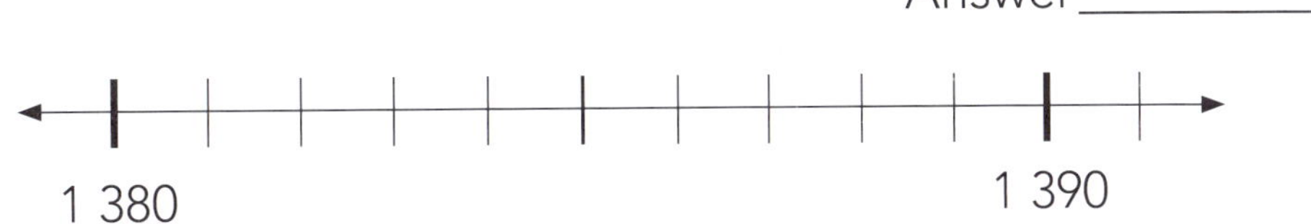

4 986 Answer ____________

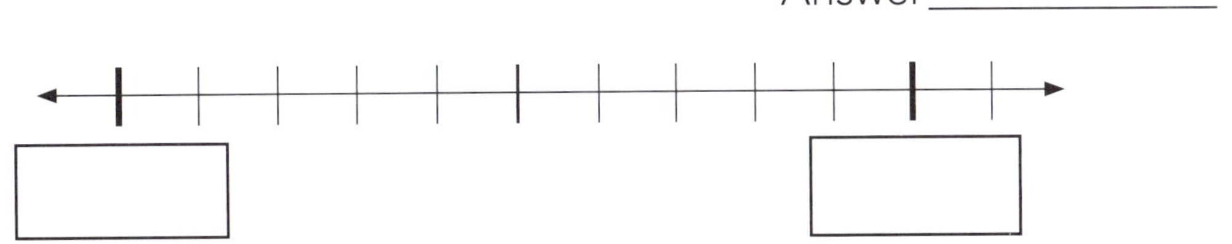

5 60 244 Answer ____________

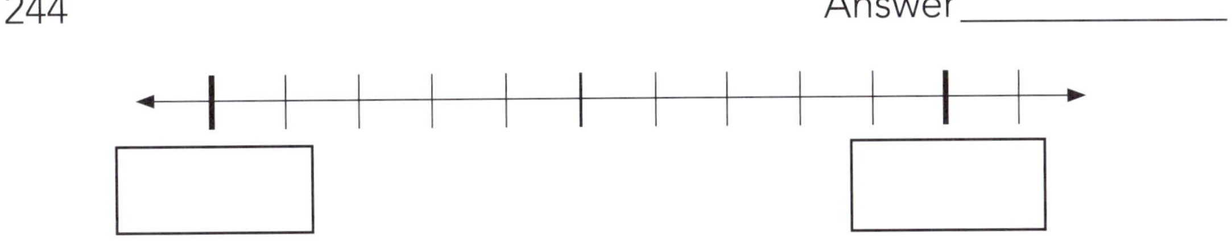

Highlight the **tens** digit, and then round these numbers to the nearest ten.

6 57 ____________ **7** 13 ____________

8 99 ____________ **9** 161 ____________

10 528 ____________ **11** 1 034 ____________

12 6 175 ____________ **13** 55 682 ____________

Mark the position of the number on the number line. Use that to help you round these numbers to the nearest hundred.

14 120 Answer ____________

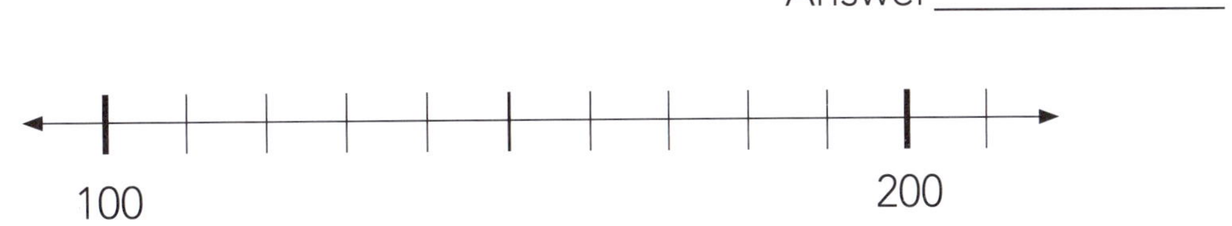

15 2 673 Answer ____________

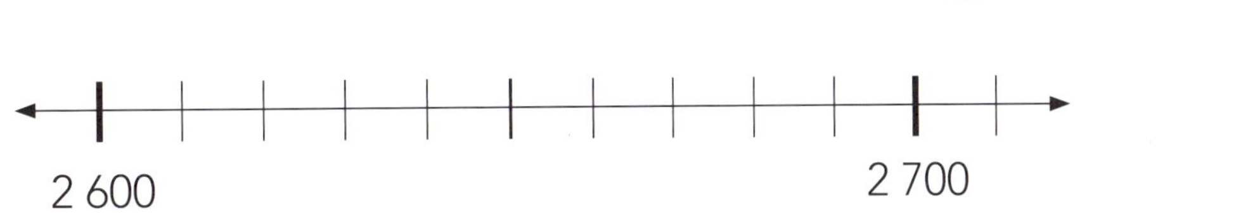

16 14 361 Answer ____________

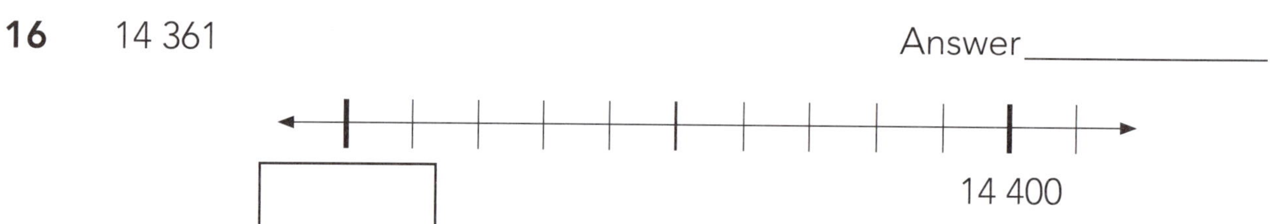

17 8 376 Answer ____________

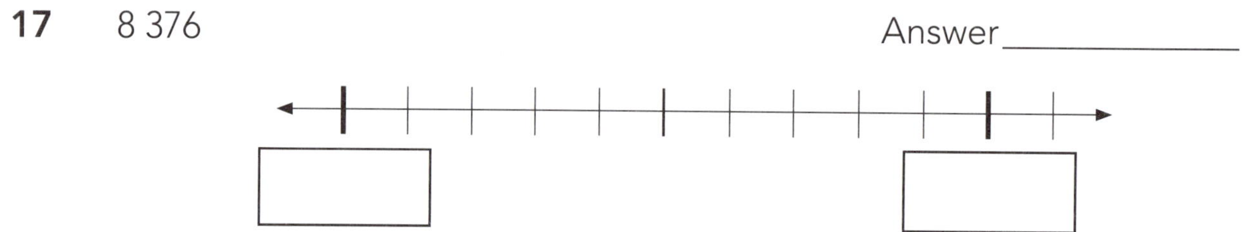

18 71 982 Answer ____________

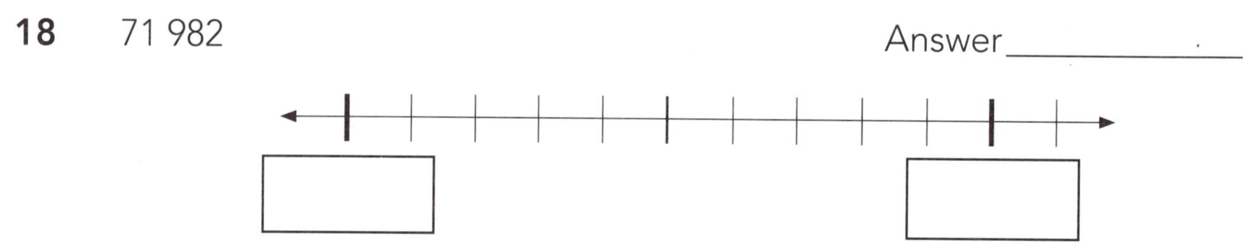

Highlight the **hundreds** digit, and then round these numbers to the nearest hundred without using a number line.

19 386 ____________ **20** 570 ____________

21 1 451 ____________ **22** 6 698 ____________

23 19 674 ____________ **24** 85 237 ____________

25 Write down what these numbers have been rounded to.

Original number	Rounded number	Rounded to the nearest:
135	140	ten
84 918	84 900	
9 451	9 000	
178 679	178 700	
62 455	62 460	
40 012 319	40 012 300	

 ISBN: 9780170447331

Rounding decimals

- The number of decimal places is the number of digits after the decimal point.

Examples:

Notice that a **0** at the **end** counts as a decimal place.

Number	9	9.2	9.2**0**	9.256	9.256**0**
Number of decimal places	0	1	2	3	4

1 Write the number of decimal places in each of the following.

	Number of decimal places		Number of decimal places
3.568	3	11.09	
1 324.5		372.0	
412.530		0.006	
0.308		5	
9.4000		42.001	

Rounding to 1 dp

- When asked to round to 1 dp (one decimal place), there should be **exactly** one digit after the decimal point.
- The process is similar to rounding to whole numbers.

Locate the **first** decimal place (**tenths**). Is the digit to its **right** 5 or more?

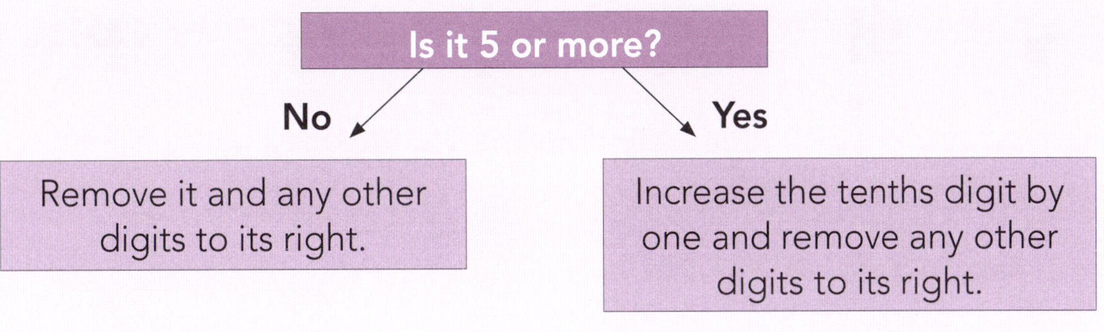

ISBN: 9780170447331

Examples: 1 Round 13.**8**4 to 1 dp.

Is 13.**8**4 closer to 13.8 or 13.9?

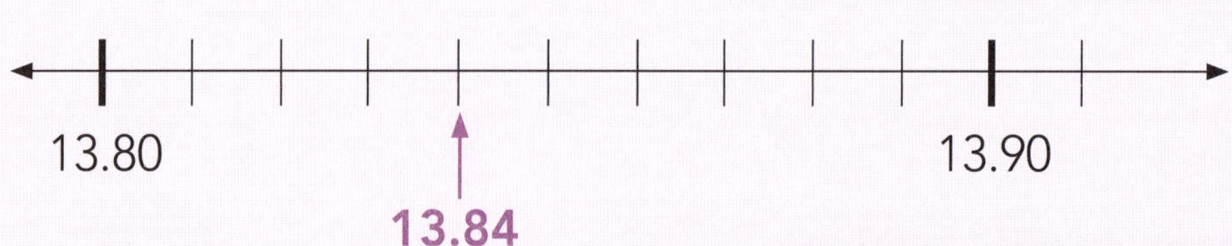

The digit to the right of the **tenths** digit (**8**) in 13.**8**4 is the **4**.
So 13.84 rounded to 1 dp is 13.8.

2 Round 63.**7**501 to 1 dp.

Is 63.**7**501 closer to 63.7 or 63.8?

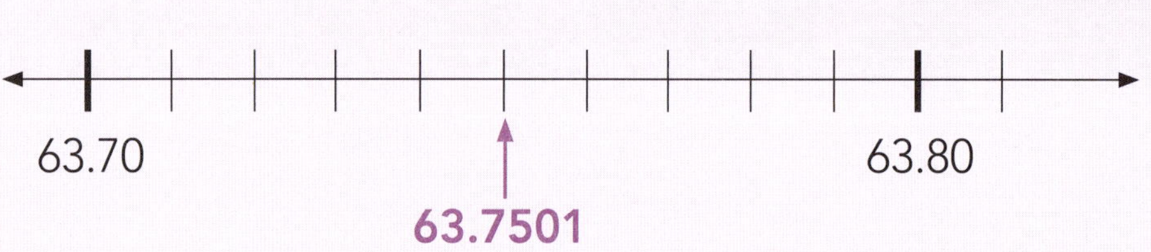

The digit to the right of the **tenths** digit (**7**) in 63.**7**501 is the **5**.
So 63.7501 rounded to 1 dp is 63.8.

3 Round 436.**9**5 to 1 dp.

Is 436.**9**5 closer to 436.9 or 437.0?

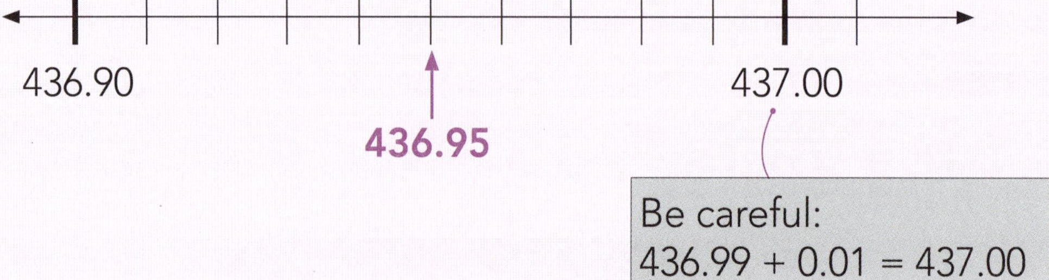

The digit to the right of the **tenths** digit (**9**) in 436.**9**5 is the **5**.
So 436.95 rounded to 1 dp is 437.0.

ISBN: 9780170447331

More examples:

	Tenths digit (purple)	Rounded to 1 dp
2.37	2.**3**7	2.4
27.592	27.**5**92	27.6
142.8325	142.**8**326	142.8
0.9613	0.**9**613	1.0

Round these numbers to 1 dp. Mark their position on the number line, and use that to help you.

2 23.47

Answer ____________

3 0.12

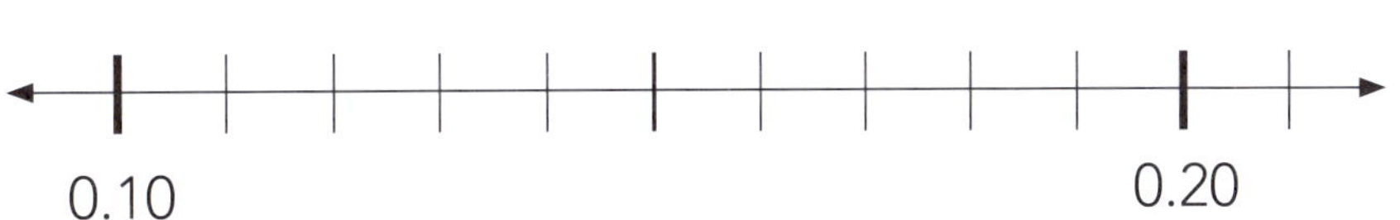

Answer ____________

4 407.09

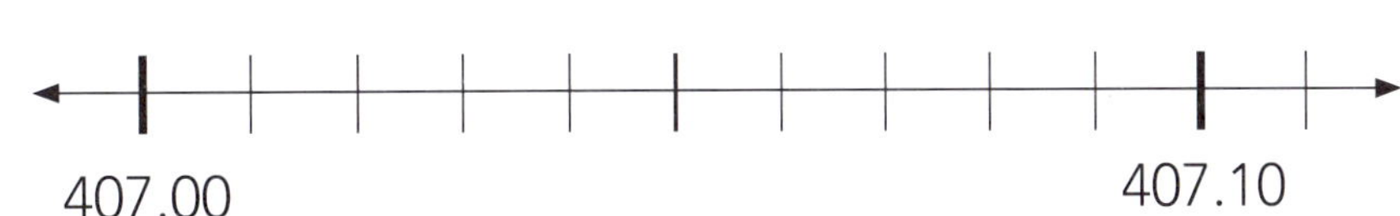

Answer ____________

5 68.45

Answer ____________

6 10.95

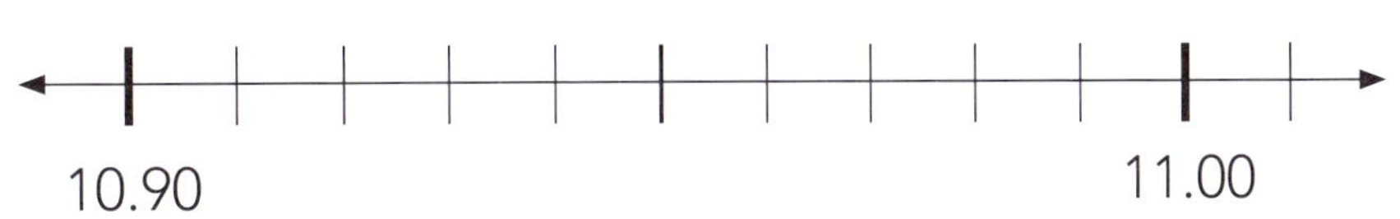

Answer ____________

ISBN: 9780170447331

7 Highlight the tenths digit, and then round these numbers to 1 dp.

	Rounded to 1 dp		Rounded to 1 dp
3.566	3.6	11.679	
24.524		372.09	
2.930		0.006	
0.308		5.98	

Answer the following questions. At the **end** of your calculations, round your answers to 1 dp.

8 Kara had 1.35 kg of carrots, 2.38 kg of potatoes and 0.42 kg of beans. What weight of vegetables did she have altogether?

9 **a** Three students are going tramping. Their food weighs a total of 5.67 kg. If they share the food equally between their packs, what weight of food should each person carry?

b Once everything, including his (rounded) share of the food, was put into Jono's pack, it weighed 13.52 kg. If the empty pack weighed 4.35 kg, what was the weight of the contents that was not food?

10 **a** The plum chutney recipe says to use 0.4 L of vinegar for every kilogram of plums. How much vinegar would be needed for 1.86 kg of plums?

b It also says that 0.25 kg of sugar is needed for a kilogram of plums. How much sugar would be needed for 1.86 kg of plums?

11 In the class relay race, there were four runners who each ran 100 m. Their times were 12.42 s, 13.15 s, 14.09 s and 12.03 s. Calculate the average time taken by each runner.

ISBN: 9780170447331

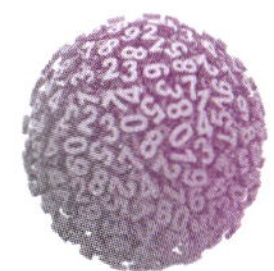

Estimations/approximations

- Sometimes an exact answer isn't necessary, so you can just estimate the answer.
- In order to estimate, round every number to **the nearest whole number**, and then do the calculations required.
- Don't forget to use BEDMAS.

Example: A list of prices:

This symbol means 'is approximately equal to'.

Oranges	$\$3.99 \approx \$4.00$
Broccoli	$\$2.29 \approx \$2.00$
Kumara	$\$2.98 \approx \$3.00$

Round each value to the nearest dollar.

The approximate total is $9.00.

$4 + 2 + 3 = 9$

Estimate the answers to the following. Do not use a calculator.

1 $5.32 + 3.89 =$ ____________

2 $4.79 + 6.11 =$ ____________

3 $19.86 - 7.46 =$ ____________

4 $10.51 - 3.31 =$ ____________

5 $28.30 + 1.13 + 4.2 =$ ____________

____

6 $11.6 - 7.3 + 1.45 =$ ____________

____

7 $3.2 \times 9.8 =$ ____________

8 $2.4 \times 4.74 =$ ____________

9 $\frac{11.9}{6.2} =$ ____________

10 $18.3 \div 1.9 =$ ____________

11 $9.43 \times 1.85 \div 2.6 =$ ____________

____

12 $8.17 \div 2.03 \times 3.33 =$ ____________

____

13 $5 \times 4.6 - 3.3 =$ ____________

____

14 $3 \times 1.85 + 4.0 =$ ____________

____

ISBN: 9780170447331

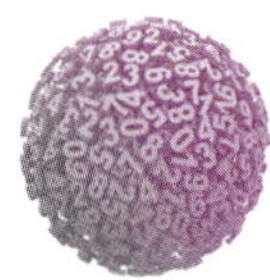

Revision 1

1 Show this calculation on the number line, and write the answer.

$1 - 3 + 7 - 2 =$ ____________

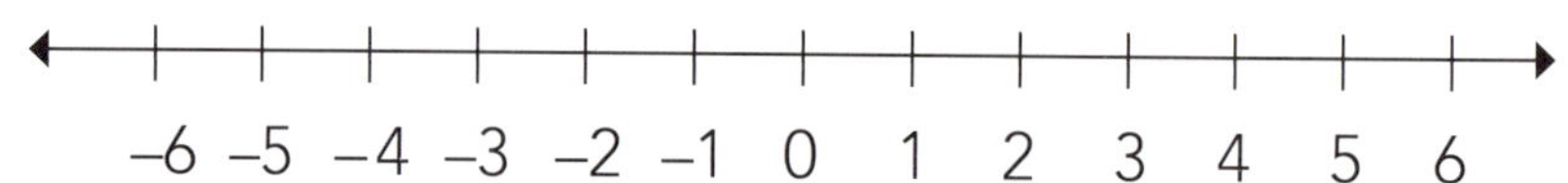

2 **a** $-1 \times -5 =$ ____________ **b** $3 \times -4 \div 6 =$ ____________

3 Highlight the numbers that belong in the boxes.

Multiples				Number	Factors			
12	4	2	5		1	4	12	30
30	1	24	3	**6**	2	3	24	36
6	36	18	28		28	5	18	6

4 What is the fourth square number? ____________

Calculate these.

5 **a** $4^3 =$ ____________ **b** $5^2 - 2^3 - 3^2 =$ ____________

6 **a** $\sqrt{25} =$ ____________ **b** $\sqrt[3]{1\,000} =$ ____________

7 **a** $5 \times 3^2 - 1 =$ ____________ **b** $20 - (6 \div 2)^2 =$ ____________

c $4 - \frac{9}{3} \times 2 =$ ____________ **d** $10 - \sqrt{12 - 3} + 4^2 =$ ____________

8 Shade the diagrams. Then write which fraction is larger and which is smaller.

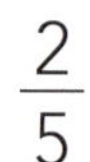

$\frac{2}{5}$

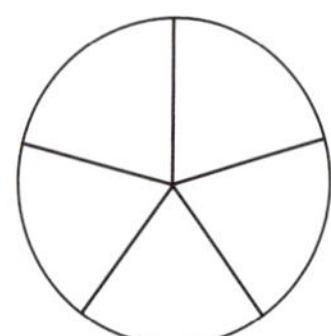

$\frac{3}{7}$

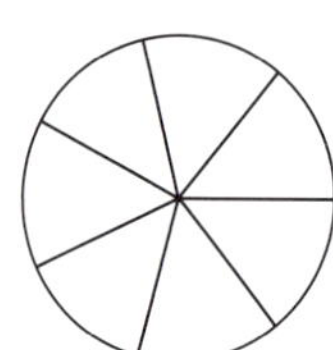

____________ ____________

 ISBN: 9780170447331

9 Simplify $\frac{18}{30}$. ______

10 Change $2\frac{1}{3}$ into an improper fraction. ______

11 Change $\frac{16}{5}$ into a mixed fraction. ______

12 Calculate these.

a $\frac{1}{5} \times \frac{3}{4} =$ ______

b $\frac{1}{7} + \frac{1}{7} + \frac{2}{7} =$ ______

c $\frac{7}{9} - \frac{3}{9} =$ ______

d $\frac{3}{4}$ of 28 = ______

13 Write down the value of the purple numeral using numerals and words.

	Number	Numerals	Words
a	561 328		
b	0.31		

14 Write these numbers using words.

a 50 067 ______

b 4.8 ______

15 Write these numbers using numerals.

a Nine hundred and two thousand, one hundred and thirty-six

______

b Seventeen hundredths ______

16 **a** 431.7 increased by two tens = ______

b 0.68 increased by five hundredths = ______

17 Write the missing decimals on the number line.

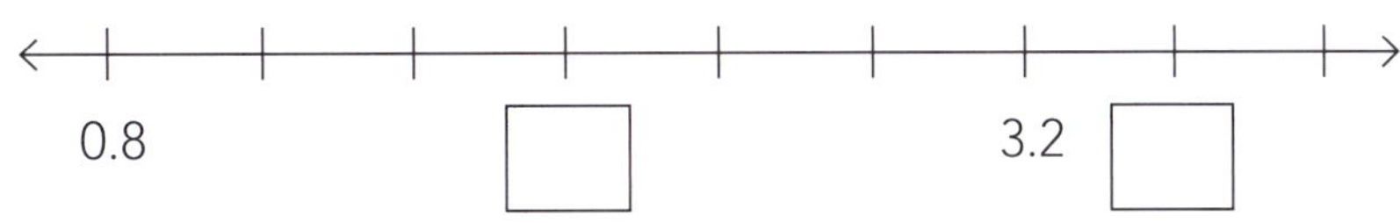

ISBN: 9780170447331

18 **a** $1.04 \times 10^3 =$ __________ **b** $0.03 \div 10^1 =$ __________

19

Percentage **dark purple** ______

Percentage light purple ______

Percentage white ______

Total: ______

20 Complete the table.

Fraction	Decimal	Percentage
$\frac{2}{5}$		
	0.75	
		15%

21 Grandpa said that he would add 5% to the amount that Rawiri saved by working during the holidays. Rawiri saved $324. How much will Grandpa add to his savings?

____

22 Grandma only sleeps for 6 hours each night. Calculate the percentage of her time she spends sleeping.

____

23 Complete the table.

	Round to the nearest:	Highlight the digit you have to round to	Answer
2 387 164	thousand		
75 016	ten		
62.95	1 dp		

 ISBN: 9780170447331

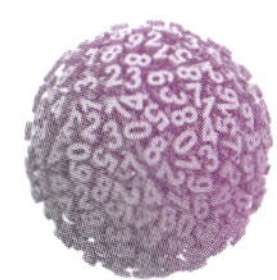

Revision 2

1 Show this calculation on the number line, and write the answer.

$-4 + 6 + 3 - 5 =$ ____________

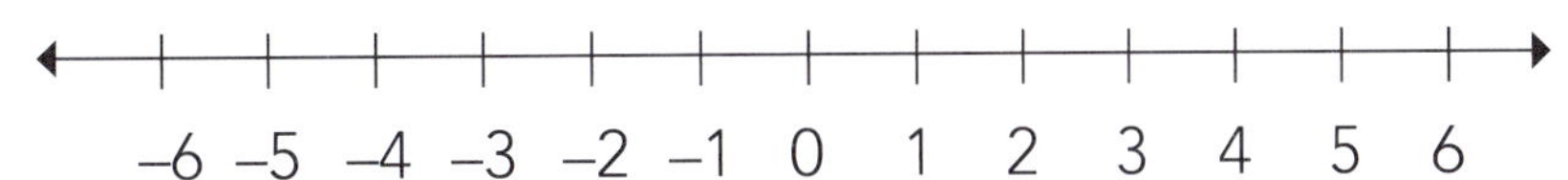

2 **a** $3 \times -11 =$ ____________ **b** $3 + -2 \times -5 =$ ____________

3 Highlight the numbers that belong in the boxes.

Multiples				Number	Factors			
20	4	30	2		20	4	30	2
10	1	24	40	**20**	10	1	24	40
100	36	5	28		100	36	5	28

4 What is the sixth square number? ____________

Calculate these.

5 **a** $2^3 =$ ____________ **b** $5^3 + 2^2 - 3^2 =$ ____________

6 **a** $\sqrt{36} =$ ____________ **b** $\sqrt[3]{27} =$ ____________

7 **a** $4 \times 2^3 + 7 - 1 =$ ____________ **b** $9 - (5 - 2)^2 =$ ____________

c $(10 - 8)^3 \div 2 - 5 =$ ____________ **d** $\sqrt{25 - 9} + 10^2 - 5 =$ ____________

8 Shade the diagrams. Then write which fraction is larger and which is smaller.

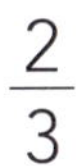

$\frac{7}{10}$

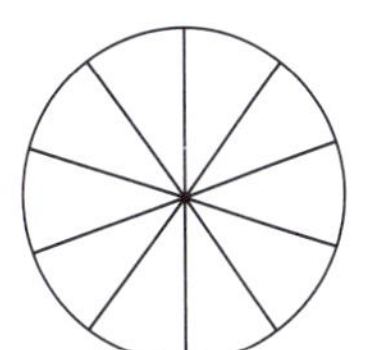

____________ ____________

ISBN: 9780170447331

9 Simplify $\frac{12}{18}$. ____________

10 Change $3\frac{1}{4}$ into an improper fraction. ____________

11 Change $\frac{23}{4}$ into a mixed fraction. ____________

12 Calculate these.

a $\frac{2}{7} \times \frac{1}{3} =$ ____________ **b** $\frac{1}{5} + \frac{3}{5} =$ ____________

c $\frac{11}{12} - \frac{7}{12} =$ ____________ **d** $\frac{5}{6}$ of 24 = ____________

13 Write down the value of the purple numeral using numerals and words.

	Number	Numerals	Words
a	359 400		
b	0.87		

14 Write these numbers using words.

a 104 804 ____________

b 7.06 ____________

15 Write these numbers using numerals.

a One hundred and twenty thousand, four hundred and five

____

b Eight hundredths ____________

16 **a** 389.71 increased by two thousandths = ____________

b 0.41 increased by nine tenths = ____________

17 Write the missing decimals on the number line.

[ ] 0.5 [ ] 2.5

 ISBN: 9780170447331

18 **a** $37.84 \times 10^2 =$ ____________ **b** $16.25 \div 10^1 =$ ____________

19

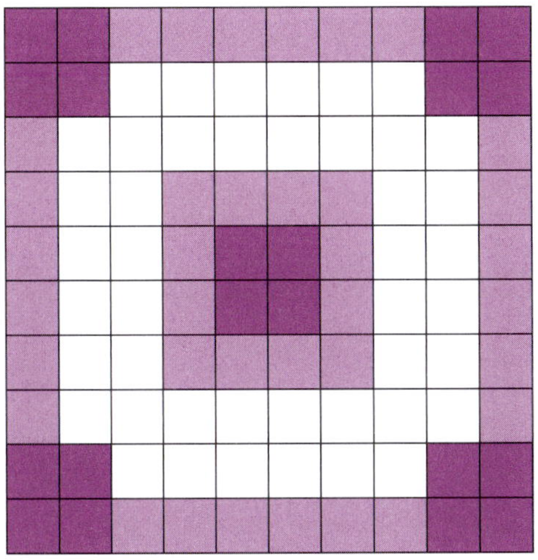

Percentage **dark purple** ______

Percentage **light purple** ______

Percentage white ______

Total: ______

20 Complete the table.

Fraction	Decimal	Percentage
$\frac{7}{10}$		
	0.45	
		75%

21 Nick has $368 in his bank account. If he earned 75% of the money by working during his holidays, how much did he earn during his holidays?

____

22 Aroha's basketball team won nine of their 15 matches during the season. What percentage of their matches did they win?

____

23 Complete the table.

	Round to the nearest:	Highlight the digit you have to round to	Answer
671 892	hundred		
36.55	whole number		
473.07	1 dp		

Answers

The language of mathematics (pp. 6–8)

Operations (p. 6)

+ add total altogether plus and together combine increase	**–** subtract decrease deduct take away minus fewer less
x double of product multiply times	**÷** goes into divide out of share between

=
gives
is
equals
will be

Words to operations (p. 7)

	Terms	Calculation	Answer
1	Twelve **shared** between three.	12 ÷ 3	4
2	What is two **more than** five?	5 + 2	7
3	Calculate three **times** two.	3 x 2	6
4	Find eight **divided** by four.	8 ÷ 4	2
5	Nine is **added** to three.	3 + 9	12
6	What is ten **take away** four?	10 – 4	6
7	What is ten **shared between** two?	10 ÷ 2	5
8	Calculate two **multiplied by** five.	2 x 5	10
9	What is one **increased by** seven?	1 + 7	8
10	Find the **difference between** five and one.	5 – 1	4
11	What is eight **plus** four?	8 + 4	12
12	Find the **sum of** three and five.	3 + 5	8
13	What is six **split between** three?	6 ÷ 3	2
14	Calculate three **less than** five.	5 – 3	2
15	What is three **minus** one?	3 – 1	2

Word questions (p. 8)

	Operator (+, –, x or ÷)	Working	Answer
1	÷	21 ÷ 3	7
2	+	23 + 14	$37
3	x	4 x 20	$80
4	÷	12 ÷ 4	3
5	–	5 – 3	2
6	+	10 + 6	16
7	÷	12 ÷ 6	2
8	x	3 x 14	42
9	–	16 – 9	7

Integers (pp. 9–17)

Adding and subtracting positive integers (pp. 9–10)

1 7

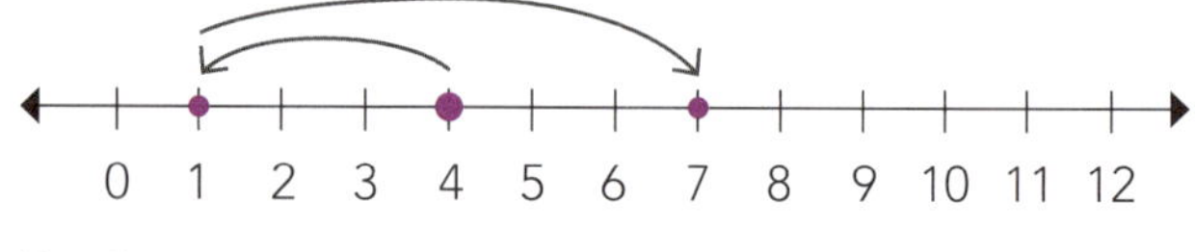

2 2

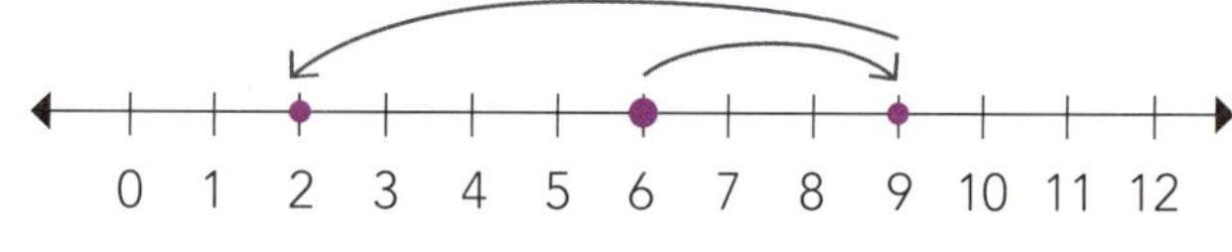

ISBN: 9780170447331

3 7

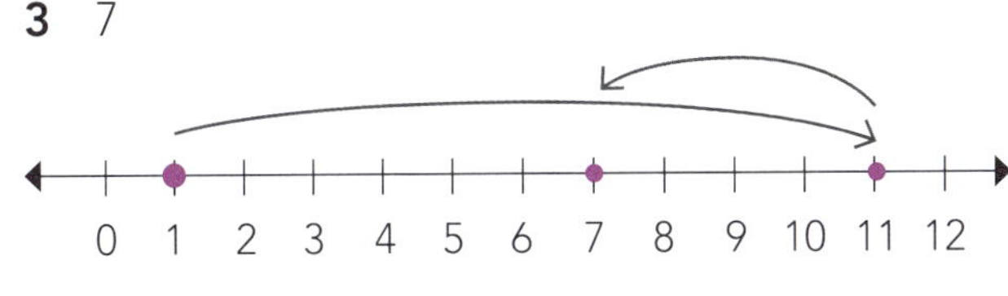

4 2

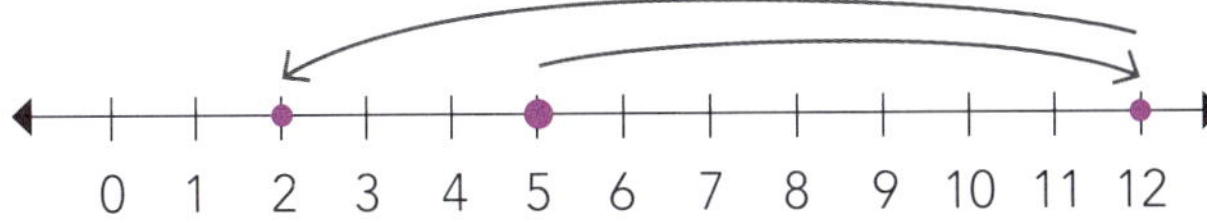

5 5

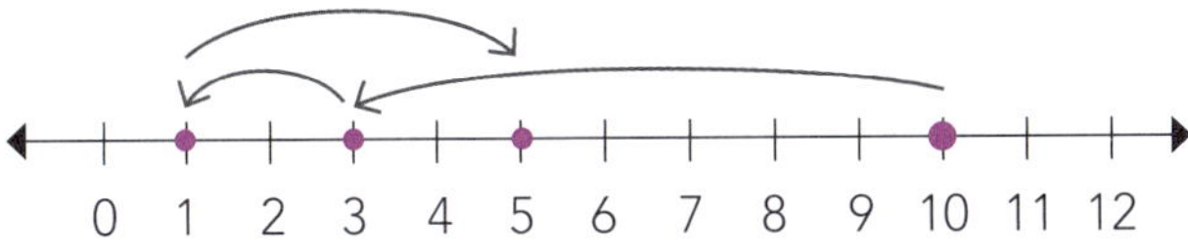

6 3

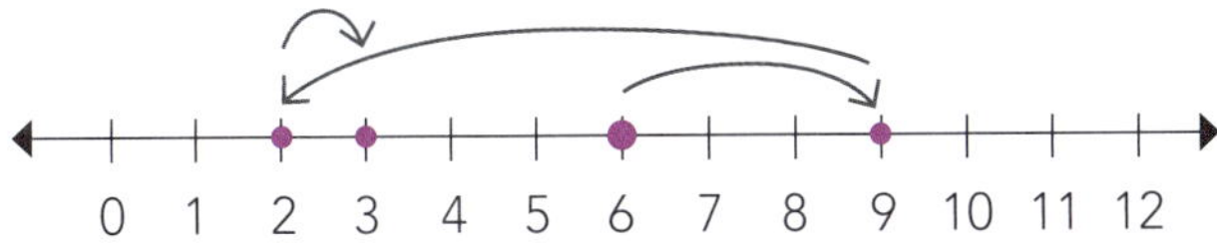

7 4
8 8
9 2
10 11
11 3
12 0
13

Smallest							Largest
0	2	3	4	6	8	10	12

14

Smallest							Largest
0	1	2	11	12	20	21	22

15

Smallest							Largest
3	4	13	14	30	34	40	43

Adding and subtracting negative integers (pp. 11–12)

1

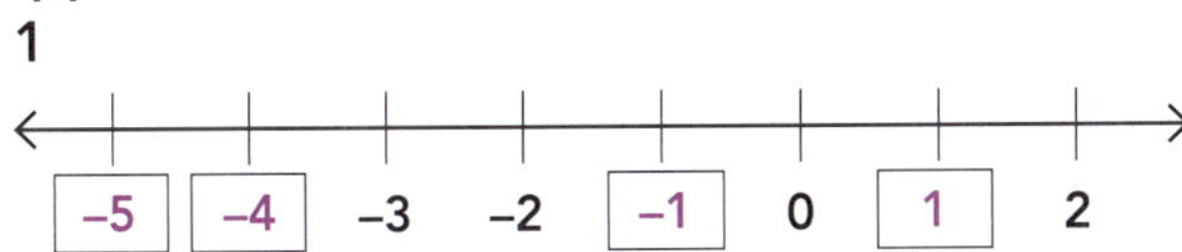

2 1

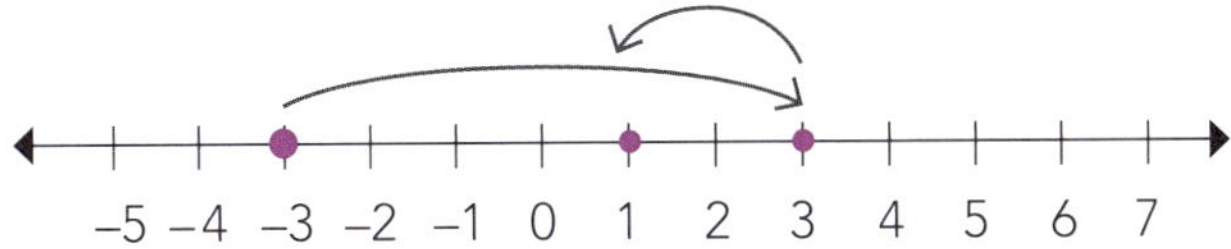

3 5

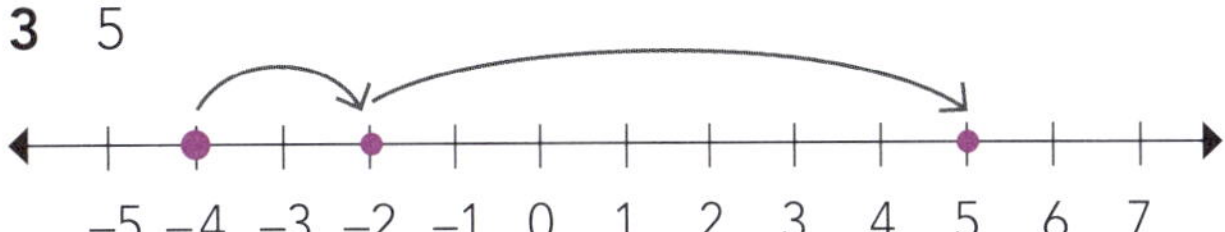

4 3

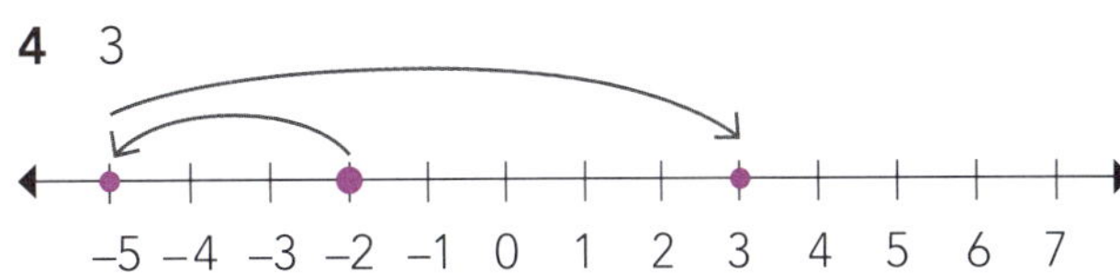

5 0

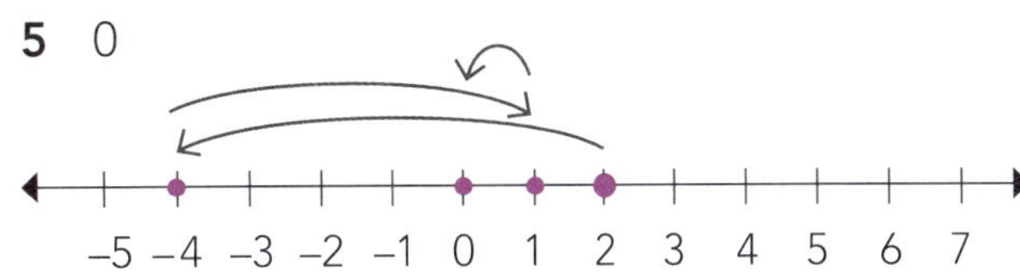

6 1

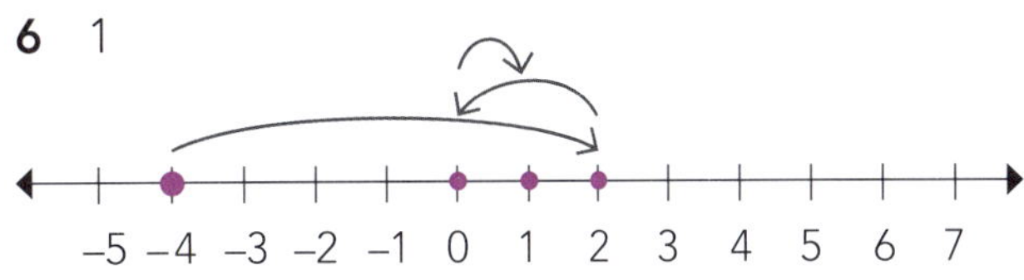

7 –3

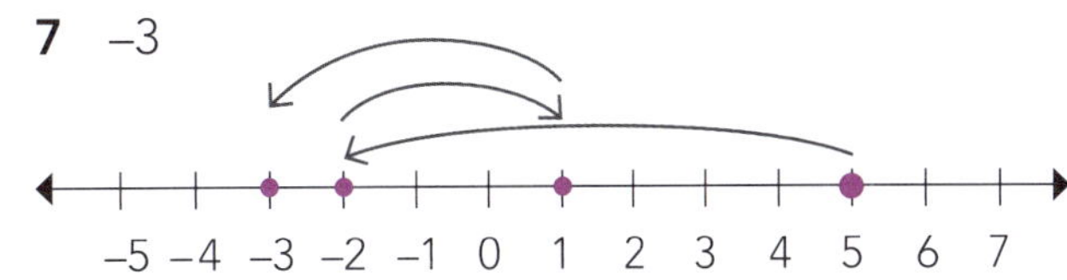

8 2
9 –1
10 –6
11 –3
12 3
13 –2
14

Smallest							Largest
–2	–1	0	1	3	4	5	6

15

Smallest							Largest
–12	–7	–2	0	4	7	9	12

16

Smallest							Largest
–31	–13	–3	0	1	3	13	31

Integers on number lines (pp. 13–15)

1 Size of the gap = 2

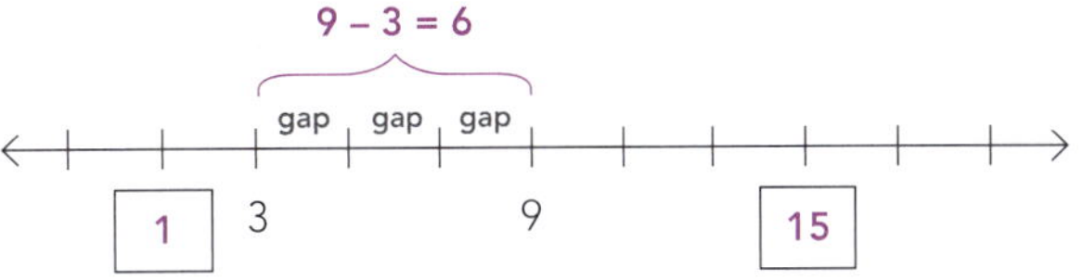

ISBN: 9780170447331

2 Size of the gap = 2

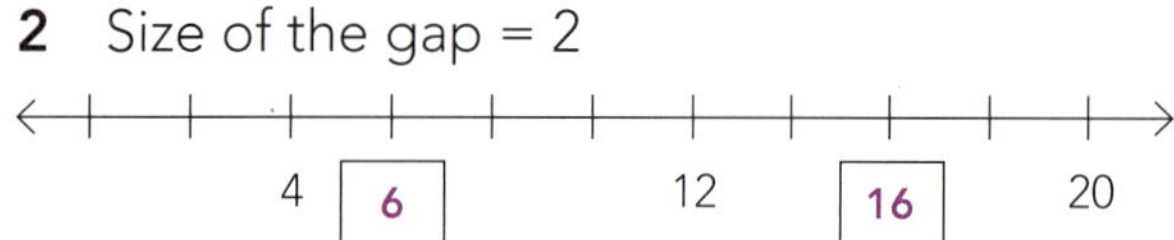

3 Size of the gap = 5

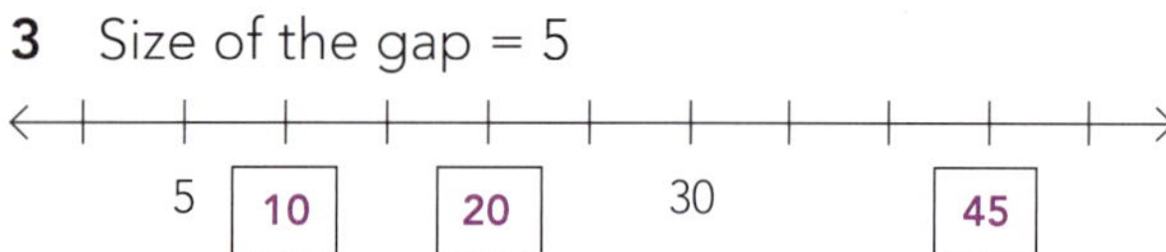

4 Size of the gap = 3

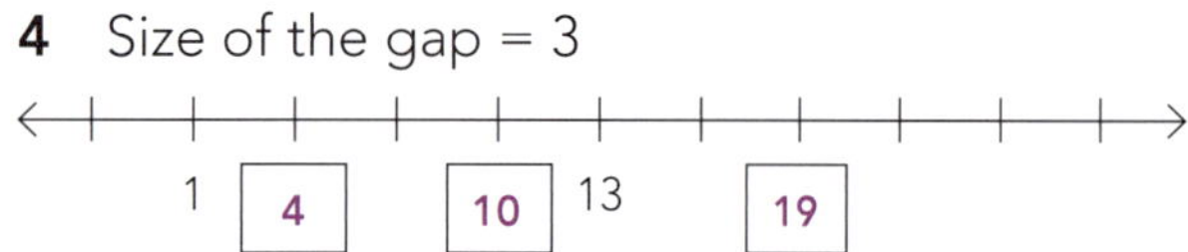

5 Size of the gap = 10

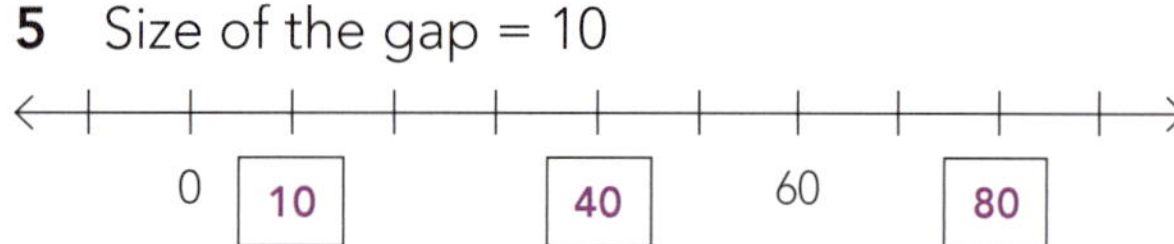

6

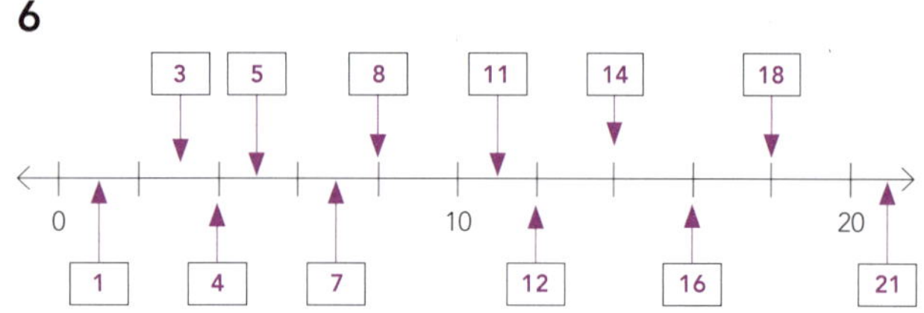

7

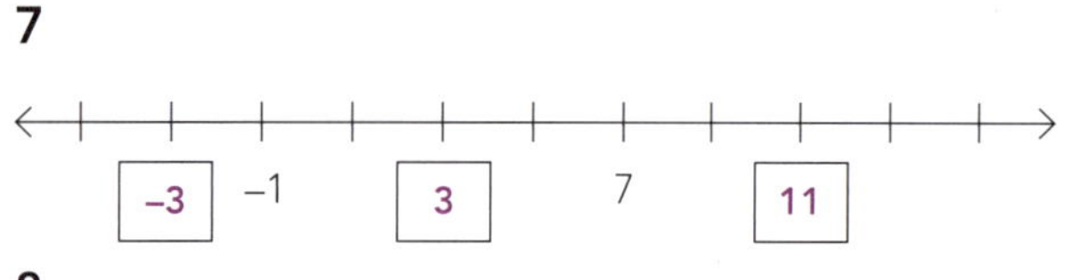

8

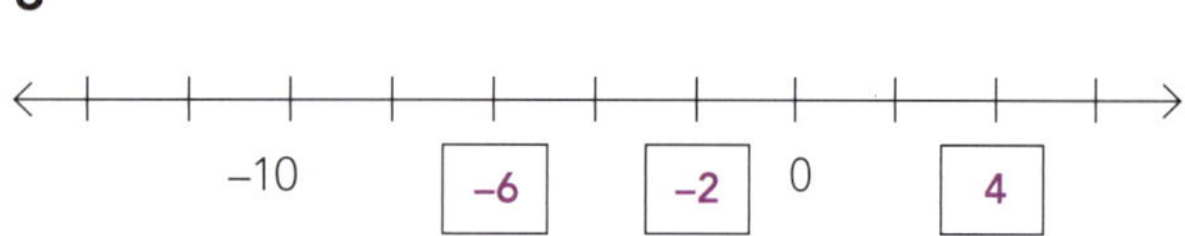

9

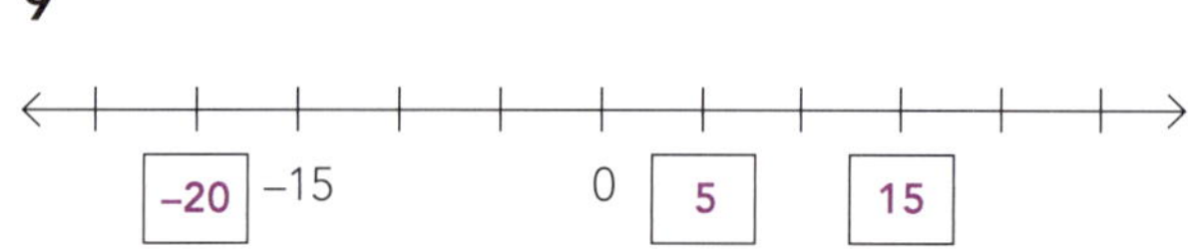

Challenge 1 (p. 15)

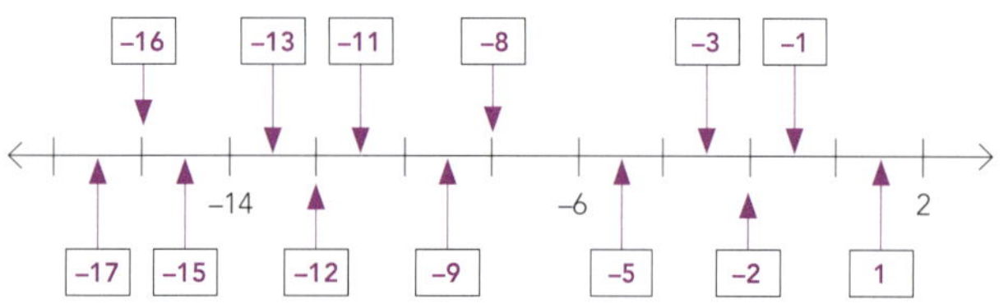

Multiplying and dividing (pp. 16–17)

1 8 **2** –5
3 –5 **4** 5
5 –6 **6** –12
7 –6 **8** 3
9 5 **10** –6
11 –16 **12** 5
13 –2 **14** 6
15 2 x **–4** = –8 **16** –4 x **3** = –12
17 **–12** ÷ –2 = 6 **18** –4 ÷ **4** = –1
19 –5 x **–3** = 15 **20** **–27** ÷ 3 = –9
21 –2 x **3** x –1 = 6 **22** 6 ÷ **3** x –2 = –4
23 There are many solutions to this. If yours is different, check it with your teacher.

–4	x	3	=	–12
10	÷	–5	=	–2
2	x	–3	=	–6

Types of numbers (pp. 18–21)

Multiples (p. 18)

1 20, 15 **2** 7, 21
3 6, 9 **4** 18, 36
5 20, 100 **6** 16, 24
7

Number	First five multiples
2	2, 4, 6, 8, 10
5	5, 10, 15, 20, 25
9	9, 18, 27, 36, 45
3	3, 6, 9, 12, 15
4	4, 8, 12, 16, 20
10	10, 20, 30, 40, 50
11	11, 22, 33, 44, 55
7	7, 14, 21, 28, 35
8	8, 16, 24, 32, 40
6	6, 12, 18, 24, 30

8

Statement	True or False
12 is a multiple of 3	True
10 is a multiple of 4	False
8 is a multiple of 2	True
18 is a multiple of 6	True
13 is a multiple of 3	False
2 is a multiple of 1	True
21 is a multiple of 7	True
15 is a multiple of 5	True
19 is a multiple of 2	False
9 is a multiple of 4	False

ISBN: 9780170447331

Factors (pp. 19–20)

1 3, 4
2 5, 2
3 1, 5
4 7, 3
5 4, 2
6 6, 2

7

Number	Factors
10	1, 2, 5, 10
8	1, 2, 4, 8
12	1, 2, 3, 4, 6, 12
9	1, 3, 9
15	1, 3, 5, 15
6	1, 2, 3, 6
18	1, 2, 3, 6, 9, 18
21	1, 3, 7, 21
20	1, 2, 4, 5, 10, 20
13	1, 13
24	1, 2, 3, 4, 6, 8, 12, 24
50	1, 2, 5, 10, 25, 50

8

Statement	True or False
2 is a factor of 15	False
1 is a factor of 7	True
5 is a factor of 10	True
3 is a factor of 12	True
6 is a factor of 18	True
2 is a factor of 1	False
7 is a factor of 20	False
4 is a factor of 12	True
3 is a factor of 13	False
4 is a factor of 20	True
8 is a factor of 18	False
1 is a factor of 15	True

9

Multiples			Number	Factors		
12	2	8		**1**	**3**	16
42	**24**		12	9	44	**12**
19	**36**	28		14	32	**4**
10	5	45		4	**5**	**10**
25	**40**		10	50	25	
30	28	2		**2**	**1**	3
16	**40**	**48**		10	6	**8**
20	12	4	8	**1**	12	**4**
24	2	26		3	**2**	20
32	1	6		2	6	**3**
24	**27**	**36**	9	18	20	**9**
18	**9**	3		**1**	4	27
55	1	2		60	45	**1**
15	5	**30**	15	**15**	**5**	8
20	**45**	3		30	7	**3**

Square numbers (p. 21)

1

Square number	Picture	Notation
1	•	1^2
4	• • • •	2^2
9	• • • • • • • • •	3^2
16	• • • • • • • • • • • • • • • •	4^2
25	• •	5^2
36	• •	6^2

ISBN: 9780170447331

Powers (pp. 22–23)

1 4^2 **2** 6^3
3 5^4 **4** 9^2
5 8^1 **6** 3^3
7 2^5 **8** 1^3
9 2 x 2 x 2 = 8 **10** 3 x 3 x 3 = 27
11 6 x 6 = 36 **12** 8 x 8 = 64
13 7 x 7 x 7 = 343 **14** 4
15 5 x 5 x 5 = 125 **16** 10 x 10 x 10 = 1 000
17 13 x 13 = 169 **18** 20 x 20 x 20 = 8 000
19 113 **20** 1
21 400 **22** 25
23 8 **24** 80

Roots (p. 24)

1 2 **2** 5
3 3 **4** 4
5 6 **6** 9
7 10 **8** 7
9 2 **10** 3
11 5 **12** 8
13 6 **14** 4

Order of operations (pp. 25–28)

B	Brackets
E	Exponents
D	Division
M	Multiplication
A	Addition
S	Subtraction

1 4 **2** 3
3 7 **4** 12
5 25 **6** 11
7 3 **8** 3
9 11 **10** 9
11 10 **12** 7
13 16 **14** 4
15 5 **16** 1
17 3 **18** 3
19 13 **20** 2
21 7 **22** 6
23 4 **+** 3 **–** 2 **x** 1 = 5 4 x 3 + 2 – 1 = 13
4 – 3 + 2 x 1 = 3 4 x 3 – 2 + 1 = 11
4 + 3 x 2 – 1 = 9 4 – 3 x 2 + 1 = –1

	Working	Answer
24	**(12 + 3 – 2) x 1**	**13**
25	**(3 + 10 – 4) x 2**	18
26	3 x 4 + 8	20
27	8 ÷ 2 – 1	3
28	(9 – 3) x 2	12
29	(11 – 5) ÷ 2	3
30	(1 x 9) ÷ 3	3
31	(12 ÷ 3) x 2 + 1	9
32	(10 – 6) x 3 – 2	10
33	(16 ÷ 4) ÷ 2 + 12	14

Using your calculator (p. 28)

A trip to the zoo

Zoe and **Lillie** were at a **loss**. They were at the zoo but couldn't find **Eloise** anywhere. They were supposed to meet her back by the **ibises** but they couldn't **see** her. **Ollie** came round the corner and shook his head — still no sign of **Eloise**. They couldn't **lie**. They were scared. They felt **ill**. Then **Zoe** heard a familiar **giggle**. **Eloise** came trotting round the corner. **Zoe** let out a **sob** and shouted at **Lillie**. Thank goodness, they wouldn't have to tell Mum.

Fractions (pp. 29–49)

The language of fractions (p. 29)

Fraction	Word	Meaning
$\frac{1}{2}$	half	÷ 2
$\frac{1}{4}$	quarter	÷ 4
$\frac{1}{3}$	third	÷ 3
$\frac{1}{5}$	fifth	÷ 5
$\frac{1}{6}$	sixth	÷ 6
$\frac{1}{7}$	seventh	÷ 7
$\frac{1}{8}$	eighth	÷ 8
$\frac{1}{9}$	ninth	÷ 9

ISBN: 9780170447331

Shading fractions (pp. 30–32)

1 Shaded $\frac{2}{3}$ Not shaded $\frac{1}{3}$

2 Shaded $\frac{3}{4}$ Not shaded $\frac{1}{4}$

3 Shaded $\frac{2}{6}$ Not shaded $\frac{4}{6}$

4 Shaded $\frac{2}{4}$ Not shaded $\frac{2}{4}$

5 Shaded $\frac{4}{5}$ Not shaded $\frac{1}{5}$

6 Shaded $\frac{3}{4}$ Not shaded $\frac{1}{4}$

7 Shaded $\frac{4}{8}$ Not shaded $\frac{4}{8}$

8 Shaded $\frac{3}{6}$ Not shaded $\frac{3}{6}$

9 Shaded $\frac{3}{5}$ Not shaded $\frac{2}{5}$

10 Shaded $\frac{5}{6}$ Not shaded $\frac{1}{6}$

11 Shaded $\frac{4}{7}$ Not shaded $\frac{3}{7}$

12 Shaded $\frac{5}{8}$ Not shaded $\frac{3}{8}$

13 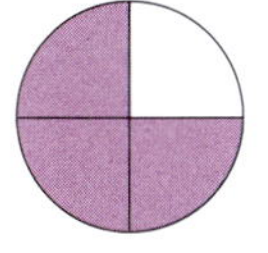

14 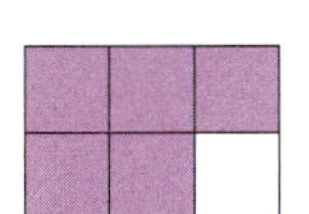

15 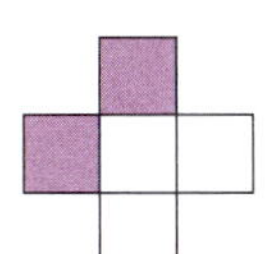

16 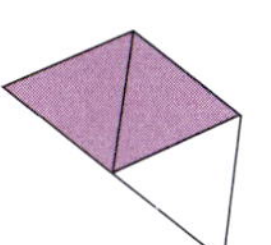

17 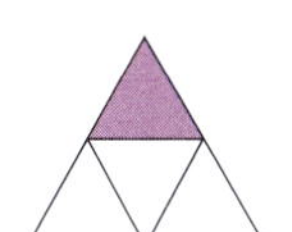

18 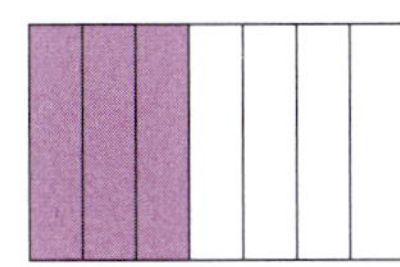

19 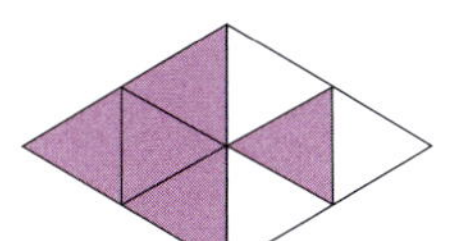

20 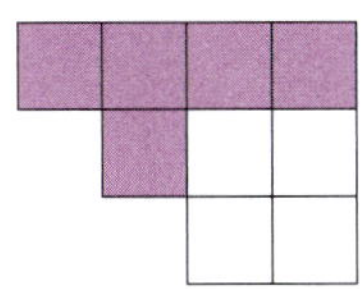

21 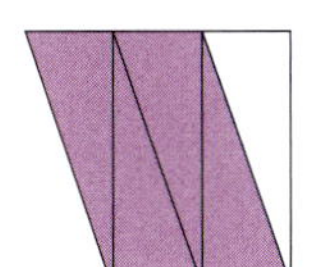

22 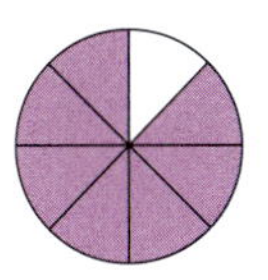

23 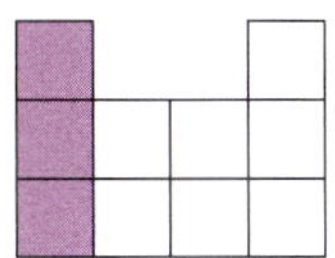

24 

Numerators and denominators (pp. 33–37)

1

Fraction

$\frac{1}{5}$

$\frac{2}{5}$

$\frac{3}{5}$

$\frac{4}{5}$

$\frac{5}{5}$

2 When the numerator gets bigger, we shade **more** of the bar.

3 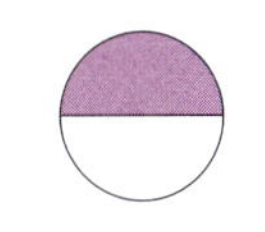

$\frac{1}{2}$ $\frac{2}{2}$

4 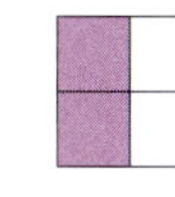 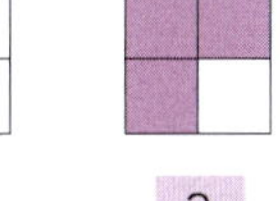

$\frac{2}{4}$ $\frac{3}{4}$

5 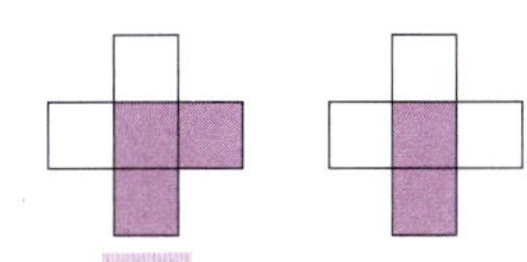

$\frac{3}{5}$ $\frac{2}{5}$

6 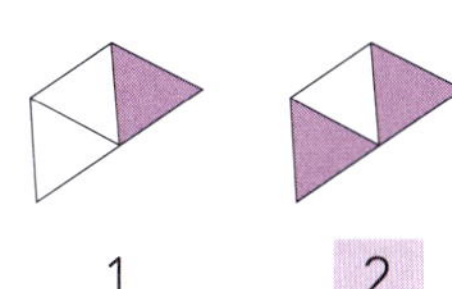

$\frac{1}{3}$ $\frac{2}{3}$

7 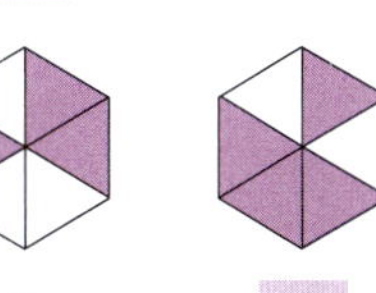

$\frac{3}{6}$ $\frac{4}{6}$

8 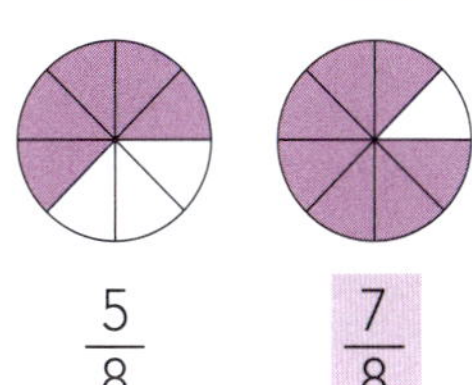

$\frac{5}{8}$ $\frac{7}{8}$

9 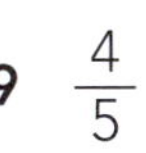 $\frac{4}{5}$

10 $\frac{5}{7}$

11 $\frac{4}{4}$

12 $\frac{7}{10}$

13 $\frac{8}{8}$

14 $\frac{3}{4}$

15

Fraction

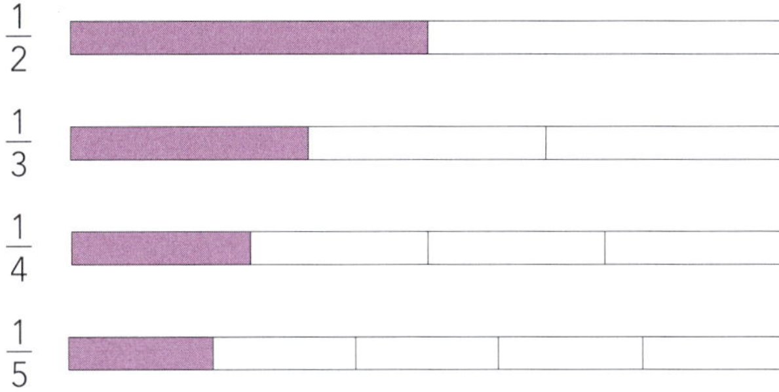

ISBN: 9780170447331

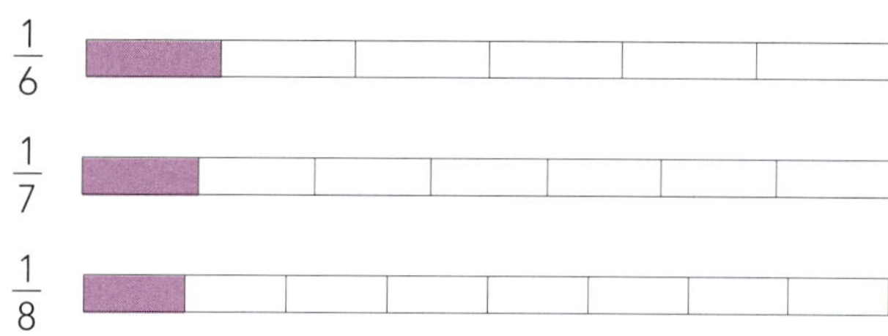

16 When the denominator gets bigger, we shade **less** of the bar.

17 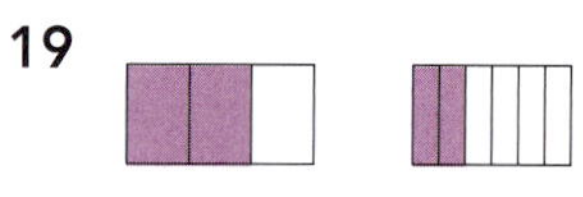

$\frac{1}{5}$ $\frac{1}{4}$

18 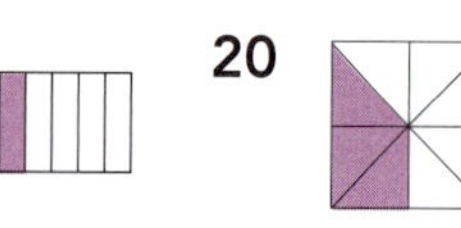 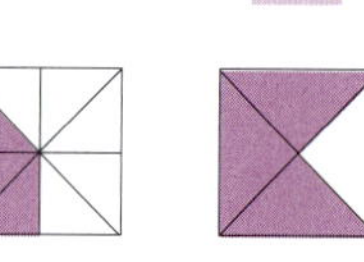

$\frac{1}{6}$ $\frac{1}{4}$

19 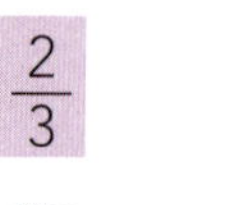 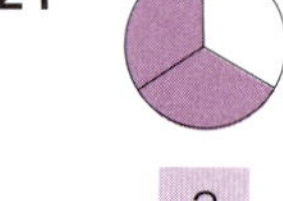

$\frac{2}{3}$ $\frac{2}{6}$

20 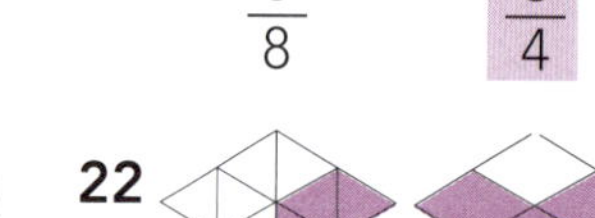

$\frac{3}{8}$ $\frac{3}{4}$

21 

$\frac{2}{3}$ $\frac{2}{5}$

22 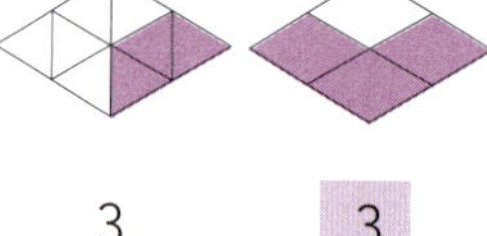

$\frac{3}{8}$ $\frac{3}{4}$

23 $\frac{1}{3}$

24 $\frac{5}{6}$

25 $\frac{2}{5}$

25 $\frac{4}{7}$

27 $\frac{1}{6}$

28 $\frac{2}{2}$

29 a

$\frac{1}{5}$ $\frac{3}{5}$ ~~$\frac{2}{5}$~~ $\frac{5}{5}$ $\frac{4}{5}$

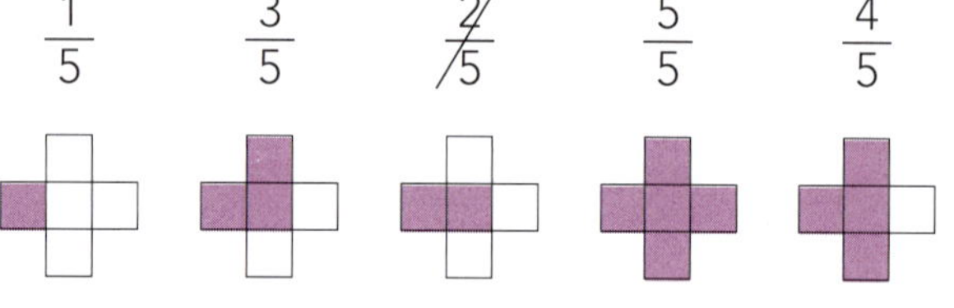

b

Smallest	$\frac{1}{5}$	$\frac{2}{5}$	$\frac{3}{5}$	$\frac{4}{5}$	$\frac{5}{5}$	Largest

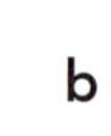

30 a

$\frac{1}{6}$ $\frac{1}{3}$ $\frac{1}{2}$ ~~$\frac{1}{4}$~~ $\frac{1}{5}$

 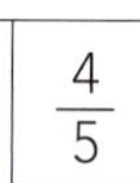

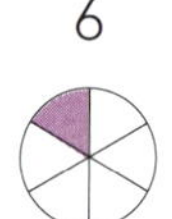 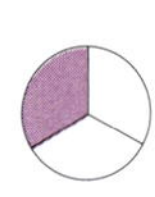 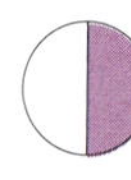 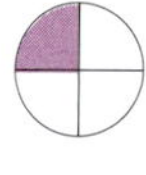 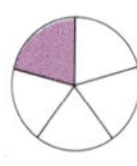 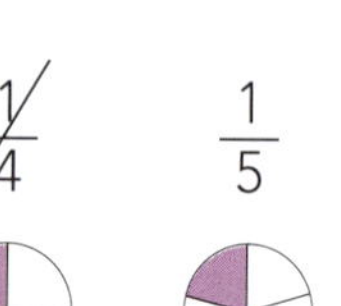

b

Smallest	$\frac{1}{6}$	$\frac{1}{5}$	$\frac{1}{4}$	$\frac{1}{3}$	$\frac{1}{2}$	Largest

What's my fraction?

$\frac{5}{7}$	$\frac{4}{11}$	$\frac{4}{10}$
$\frac{2}{17}$	$\frac{1}{3}$	$\frac{8}{13}$

Equivalent fractions (p. 38)

1 $\frac{2}{6}$ $\frac{1}{3}$

2 $\frac{6}{8}$ $\frac{3}{4}$

3 $\frac{2}{3}$ $\frac{6}{9}$

4 $\frac{4}{10}$ $\frac{2}{5}$

Simplifying fractions (pp. 39–40)

1 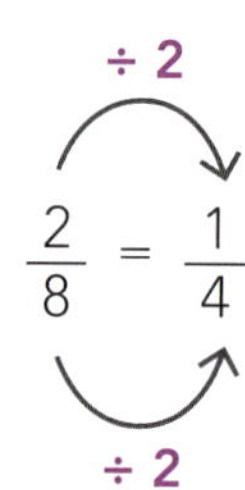

$\frac{2}{8} = \frac{1}{4}$ (÷ 2, ÷ 2)

2 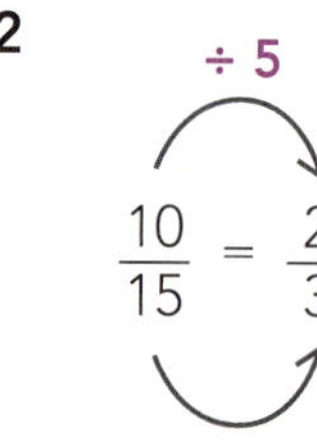

$\frac{10}{15} = \frac{2}{3}$ (÷ 5, ÷ 5)

3 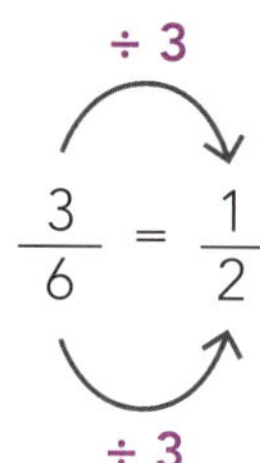

$\frac{3}{6} = \frac{1}{2}$ (÷ 3, ÷ 3)

4 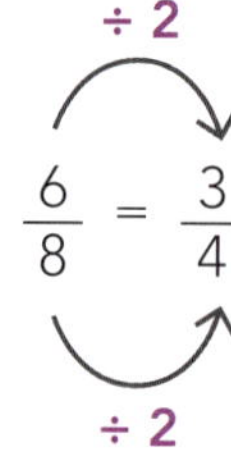

$\frac{6}{8} = \frac{3}{4}$ (÷ 2, ÷ 2)

5 $\frac{5}{10} = \frac{1}{2}$ (÷ 5, ÷ 5)

6 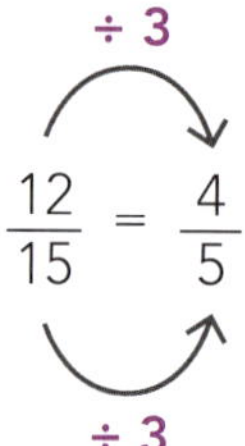

$\frac{12}{15} = \frac{4}{5}$ (÷ 3, ÷ 3)

7 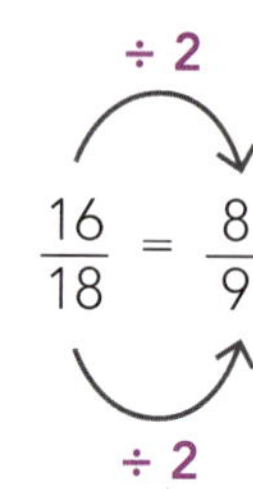

$\frac{16}{18} = \frac{8}{9}$ (÷ 2, ÷ 2)

 ISBN: 9780170447331

8

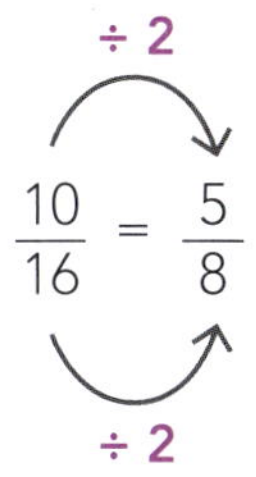

$$\frac{10}{16} \overset{\div 2}{=} \frac{5}{8}$$

9

$$\frac{9}{12} \overset{\div 3}{=} \frac{3}{4}$$

10 $\frac{20}{25} = \frac{4}{5}$

11 $\frac{21}{27} = \frac{7}{9}$

12 $\frac{10}{20} = \frac{5}{10} = \frac{1}{2} = \frac{8}{16} = \frac{4}{8} = \frac{3}{6} = \frac{6}{12}$

Comparing fractions (pp. 41–42)

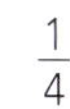

1 $\frac{2}{6}$ $\frac{1}{4}$

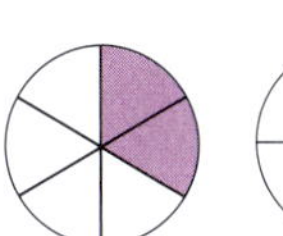

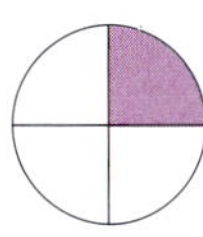
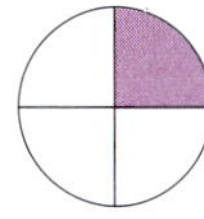
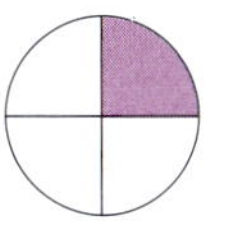

Larger Smaller

2 $\frac{5}{7}$ $\frac{3}{5}$

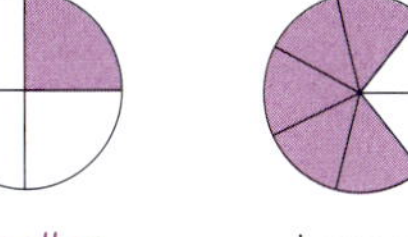

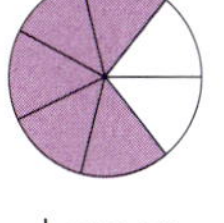

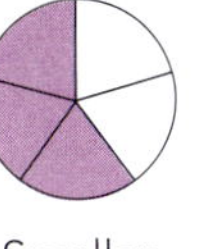

Larger Smaller

3 $\frac{2}{5}$ $\frac{1}{3}$

Larger Smaller

4 $\frac{4}{9}$ $\frac{2}{4}$

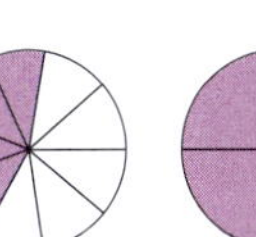

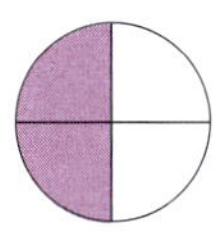

Smaller Larger

5 $\frac{1}{4}$ $\frac{2}{7}$

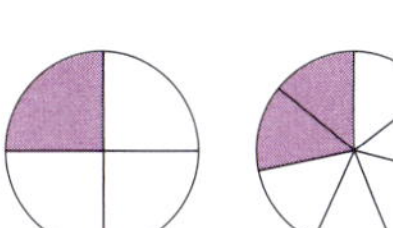
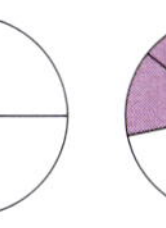
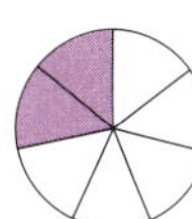

Smaller Larger

6 $\frac{5}{8}$ $\frac{3}{5}$

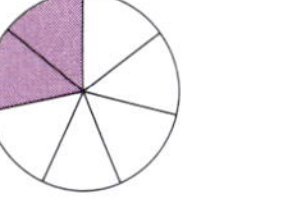
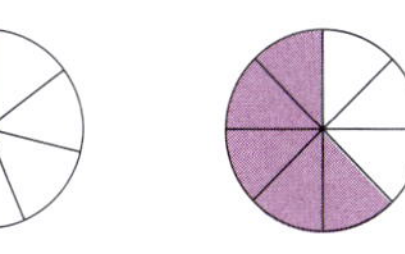
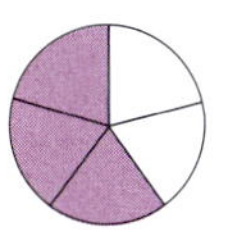

Larger Smaller

7 $\frac{4}{10}$ $\frac{1}{3}$

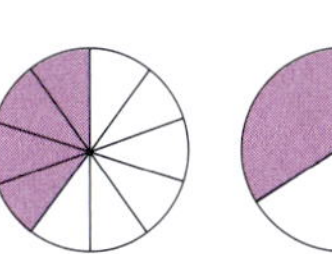
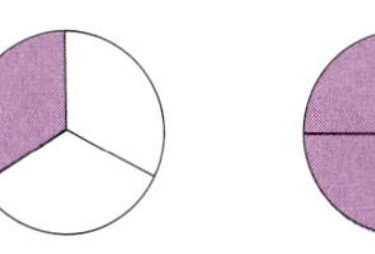

Larger Smaller

8 $\frac{3}{4}$ $\frac{4}{6}$

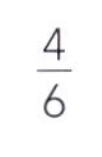
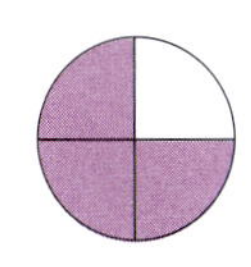
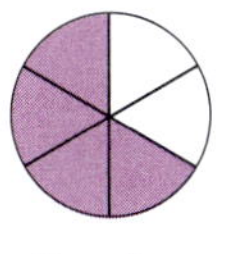

Larger Smaller

9

$\frac{4}{9}$ 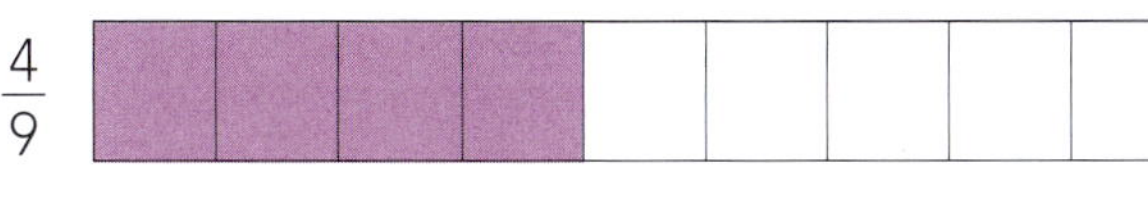

$\frac{1}{3}$ 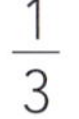

$\frac{2}{5}$ 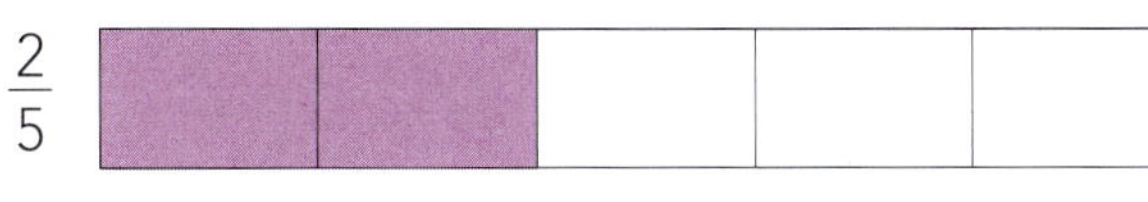

Smallest $\frac{1}{3}$ $\frac{2}{5}$ $\frac{4}{9}$ Largest

Converting between improper and mixed fractions (pp. 43–46)

	Improper fraction	Mixed fraction
1	$\frac{5}{4}$ Five quarters	$1\frac{1}{4}$ One and one quarter
2	$\frac{5}{3}$ Five thirds	$1\frac{2}{3}$ One and two thirds
3	$\frac{7}{4}$ Seven quarters	$1\frac{3}{4}$ One and three quarters
4	$\frac{11}{6}$ Eleven sixths	$1\frac{5}{6}$ One and five sixths
5	$\frac{8}{5}$ Eight fifths	$1\frac{3}{5}$ One and three fifths
6	$\frac{5}{2}$ Five halves	$2\frac{1}{2}$ Two and one half
7	$\frac{7}{3}$ Seven thirds	$2\frac{1}{3}$ Two and one third

8 $\frac{7}{4} = \frac{4}{4} + \frac{3}{4} = 1\frac{3}{4}$

9 $\frac{9}{5} = \frac{5}{5} + \frac{4}{5} = 1\frac{4}{5}$

10 $\frac{7}{6} = \frac{6}{6} + \frac{1}{6} = 1\frac{1}{6}$

11 $\frac{5}{2} = \frac{2}{2} + \frac{2}{2} + \frac{1}{2} = 2\frac{1}{2}$

12 $\frac{8}{3} = \frac{3}{3} + \frac{3}{3} + \frac{2}{3} = 2\frac{2}{3}$

ISBN: 9780170447331

13 $\frac{10}{7} = \frac{7}{7} + \frac{3}{7} = 1\frac{3}{7}$

14 $\frac{7}{2} = \frac{2}{2} + \frac{2}{2} + \frac{2}{2} + \frac{1}{2} = 3\frac{1}{2}$

15 $\frac{9}{4} = \frac{4}{4} + \frac{4}{4} + \frac{1}{4} = 2\frac{1}{4}$

16 $\frac{10}{6} = \frac{6}{6} + \frac{4}{6} = 1\frac{4}{6}$

17 $\frac{11}{3} = \frac{3}{3} + \frac{3}{3} + \frac{3}{3} + \frac{2}{3} = 3\frac{2}{3}$

18 $\frac{6}{5} = \frac{5}{5} + \frac{1}{5} = 1\frac{1}{5}$

19 $\frac{13}{7} = \frac{7}{7} + \frac{6}{7} = 1\frac{6}{7}$

20 $1\frac{3}{4} = \frac{4}{4} + \frac{3}{4} = \frac{7}{4}$

21 $2\frac{1}{5} = \frac{5}{5} + \frac{5}{5} + \frac{1}{5} = \frac{11}{5}$

22 $1\frac{2}{6} = \frac{6}{6} + \frac{2}{6} = \frac{8}{6}$

23 $2\frac{2}{3} = \frac{3}{3} + \frac{3}{3} + \frac{2}{3} = \frac{8}{3}$

24 $3\frac{1}{2} = \frac{2}{2} + \frac{2}{2} + \frac{2}{2} + \frac{1}{2} = \frac{7}{2}$

25 $1\frac{3}{5} = \frac{5}{5} + \frac{3}{5} = \frac{8}{5}$

26 $1\frac{1}{3} = \frac{3}{3} + \frac{1}{3} = \frac{4}{3}$

27 $1\frac{5}{9} = \frac{9}{9} + \frac{5}{9} = \frac{14}{9}$

28 $2\frac{1}{7} = \frac{7}{7} + \frac{7}{7} + \frac{1}{7} = \frac{15}{7}$

29 $2\frac{2}{4} = \frac{4}{4} + \frac{4}{4} + \frac{2}{4} = \frac{10}{4}$

30 $1\frac{1}{5} = \frac{5}{5} + \frac{1}{5} = \frac{6}{5}$

31 $2\frac{3}{8} = \frac{8}{8} + \frac{8}{8} + \frac{3}{8} = \frac{19}{8}$

Adding and subtracting fractions (p. 47)

1 $\frac{2}{4} + \frac{1}{4} = \frac{3}{4}$

2 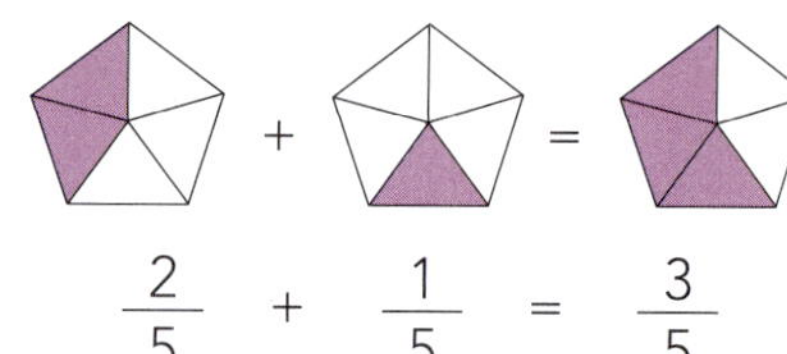

$\frac{2}{5} + \frac{1}{5} = \frac{3}{5}$

3 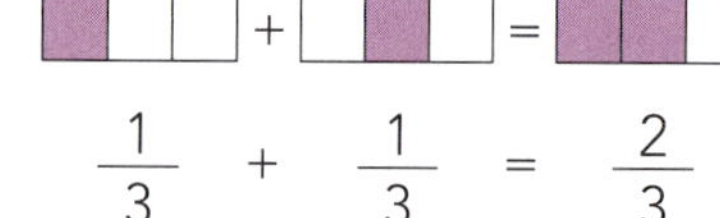

$\frac{1}{3} + \frac{1}{3} = \frac{2}{3}$

4 $\frac{3}{4} - \frac{1}{4} = \frac{2}{4}$

5 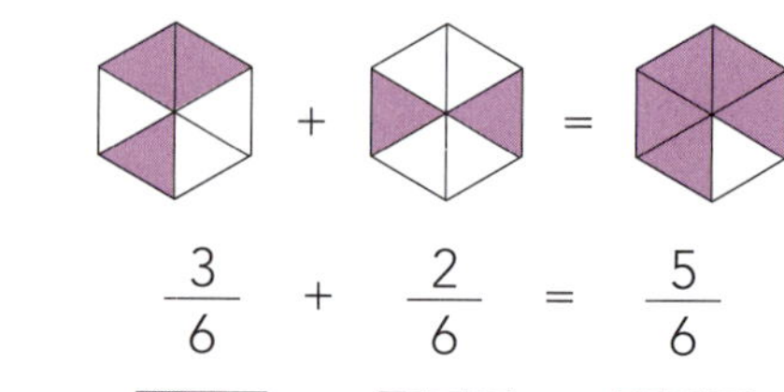

$\frac{3}{6} + \frac{2}{6} = \frac{5}{6}$

6 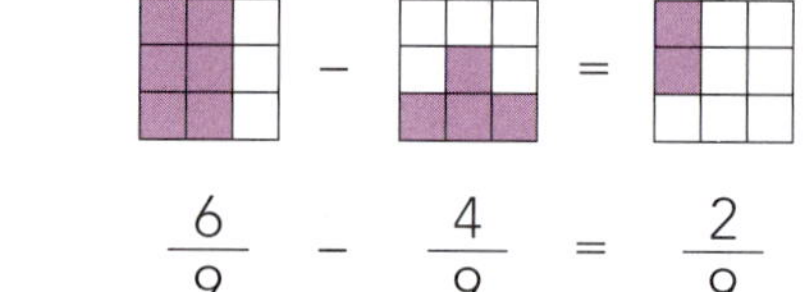

$\frac{6}{9} - \frac{4}{9} = \frac{2}{9}$

7 $\frac{3}{5} + \frac{1}{5} = \frac{4}{5}$

8 $\frac{5}{8} - \frac{3}{8} = \frac{2}{8}$

Multiplying fractions (p. 48)

1 $\frac{2}{3} \times \frac{1}{2} = \frac{2 \times 1}{3 \times 2} = \frac{2}{6} = \frac{1}{3}$

2 $\frac{2}{5} \times \frac{3}{4} = \frac{2 \times 3}{5 \times 4} = \frac{6}{20} = \frac{3}{10}$

3 $\frac{1}{12}$

4 $\frac{2}{5}$

5 $\frac{1}{20}$

6 $\frac{5}{9}$

7 $\frac{2}{7}$

8 $\frac{1}{7}$

Fractions of a quantity (p. 49)

1 $\frac{1}{3}$ of $18 = \frac{1}{3} \times \frac{18}{1} = \frac{18}{3} = 6$

2 $\frac{1}{2}$ of $42 = \frac{1}{2} \times \frac{42}{1} = \frac{42}{2} = 21$

 ISBN: 9780170447331

3 5
4 7
5 60
6 18
7 18
8 24

Decimals (pp. 50–61)

Place value (pp. 50–53)

		Number	Words
1	4 619	4 000	Four thousand
2	312 568	10 000	Ten thousand
3	7 056	50	Fifty
4	783 124	100	One hundred
5	381 266	300 000	Three hundred thousand
6	801 625	1 000	One thousand
7	8 924 119	8 000 000	Eight million
8	13 791	700	Seven hundred
9	8 320	20	Twenty

10 Two thousand, three hundred and eighty-one
11 Ninety thousand, one hundred and sixty-two
12 Five hundred thousand, four hundred and eighty-seven
13 One hundred and three thousand, two hundred
14 Six hundred and seventy thousand and fifty-four
15 Five hundred and two thousand and seven
16 One million, six hundred thousand, one hundred and fifty-one
17 400 325
18 8 068
19 34 911
20 253 406
21 2 137 000
22 500 068
23 70 413
24 0.06

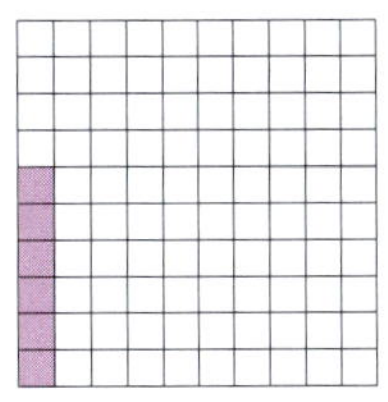

Six hundredths

25 0.2

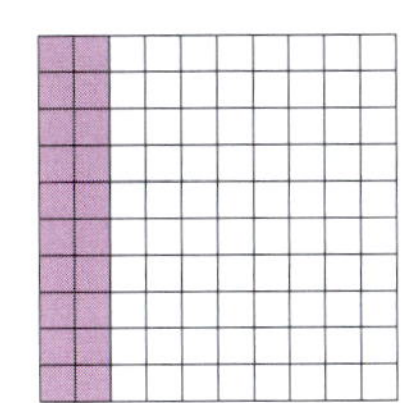

Two tenths or twenty hundredths

26 0.34

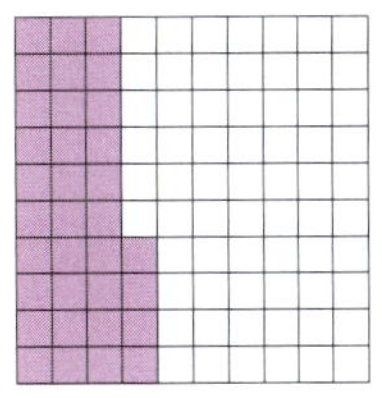

Three tenths and four hundredths or thirty-four hundredths

27 0.81

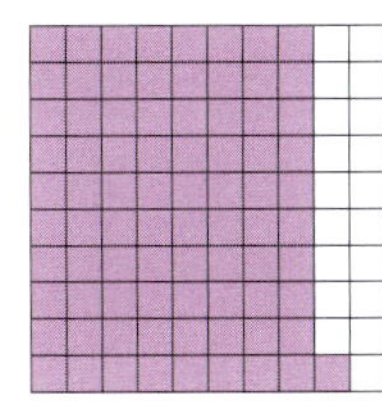

Eight tenths and one hundredth or eighty-one hundredths

		Decimal	Fraction	Words
28	0.91	0.9	$\frac{9}{10}$ $\frac{90}{100}$	Nine tenths or ninety hundredths
29	0.06	0.06	$\frac{6}{100}$	Six hundredths
30	9.54	0.04	$\frac{4}{100}$	Four hundredths
31	61.37	0.3	$\frac{3}{10}$ $\frac{30}{100}$	Three tenths or thirty hundredths
32	0.05	0.05	$\frac{5}{100}$	Five hundredths
33	8.20	0.2	$\frac{2}{10}$ $\frac{20}{100}$	Two tenths or twenty hundredths
34	13.60	0.00	$\frac{0}{100}$	Zero hundredths
35	0.01	0.01	$\frac{1}{100}$	One hundredth

36 0.9
37 0.02
38 0.16
39 1.2
40 0.83
41 0.56
42 1.09
43 2.34
44 4.21
45 2.07

Decimals on number lines (pp. 54–55)

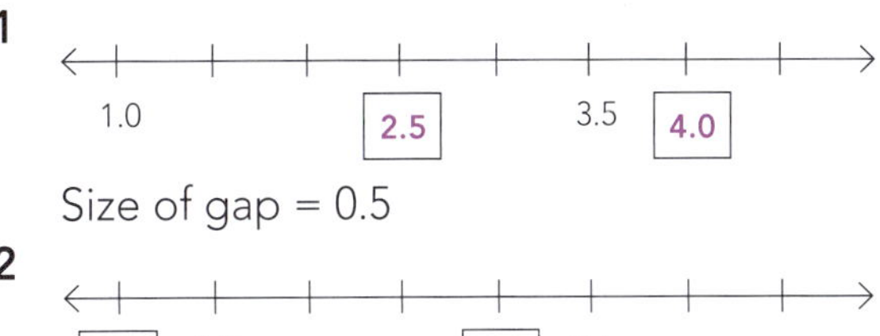

ISBN: 9780170447331

3

Size of gap = 0.2

4

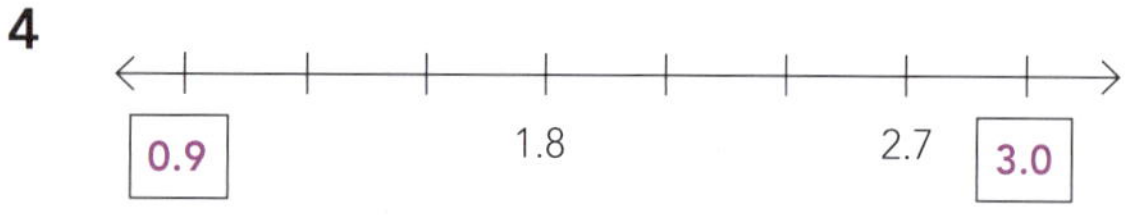

Size of gap = 0.3

5

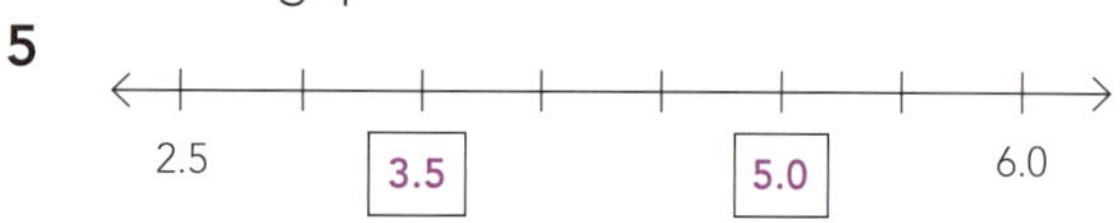

Size of gap = 0.5

6

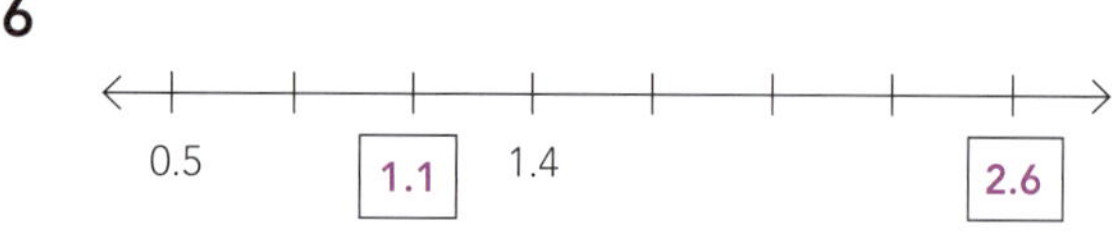

Size of gap = 0.3

7

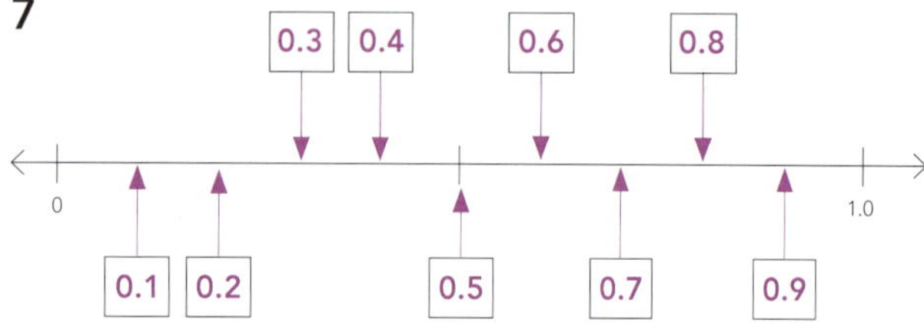

Comparing decimals (pp. 56–58)

1	8.8	**2**	1 025
3	3.49	**4**	72.07
5	123.1	**6**	11.11
7	1 010	**8**	0.10
9	11.0	**10**	5.10
11	101.0	**12**	0.21
13	88.2	**14**	41.7
15	5 215	**16**	9.5
17	30.55	**18**	1 652.3
19	654.17	**20**	735.93
21	0.05	**22**	71.0

	Smallest	Largest
23	467	764
24	2 069	9 620
25	24 569	96 542
26	3 068	8 630
27	102 468	864 210

28 3.11, 3.16, 3.61

29 0.01, 0.10, 0.11

30 2.11, 2.12, 2.21

31 0.78, 0.87, 0.88

32 0.12, 0.20, 0.21

33 0.07, 0.09, 0.11, 0.13, 0.15, 0.17
Rule: Add two hundredths (or Add 0.02)

34 3.25, 2.95, 2.65, 2.35, 2.05, 1.75
Rule: Subtract three tenths (or Subtract 0.3)

35 10.81, 10.72, 10.63, 10.54, 10.45, 10.36
Rule: Subtract nine hundredths (or Subtract 0.09)

Multiplying and dividing decimals by powers of 10 (p. 59)

1	125	**2**	780
3	8 460	**4**	6.3
5	927 500	**6**	6
7	5.32	**8**	1.775
9	5.8	**10**	41.29
11	0.006	**12**	1.73
13	0.015	**14**	1 000

Using decimals to compare fractions (pp. 60–61)

1 $\frac{1}{4} = 0.25$, $\frac{13}{50} = 0.26$ — $\frac{13}{50}$ is larger than $\frac{1}{4}$

2 $\frac{3}{20} = 0.15$, $\frac{1}{5} = 0.20$ — $\frac{1}{5}$ is larger than $\frac{3}{20}$

3 $\frac{3}{4} = 0.75$, $\frac{19}{25} = 0.76$ — $\frac{19}{25}$ is larger than $\frac{3}{4}$

4 $\frac{17}{20} = 0.85$, $\frac{4}{5} = 0.80$ — $\frac{17}{20}$ is larger than $\frac{4}{5}$

5 $\frac{11}{20} = 0.55$, $\frac{13}{25} = 0.52$ — $\frac{11}{20}$ is larger than $\frac{13}{25}$

6 $\frac{7}{25} = 0.28$, $\frac{1}{4} = 0.25$ — $\frac{7}{25}$ is larger than $\frac{1}{4}$

7 $\frac{3}{4} = 0.75$, $\frac{39}{50} = 0.78$ — $\frac{39}{50}$ is larger than $\frac{3}{4}$

8 $\frac{7}{8} = 0.875$, $\frac{43}{50} = 0.86$ — $\frac{7}{8}$ is larger than $\frac{43}{50}$

ISBN: 9780170447331

Challenge 2 (p. 61)

1 $\frac{9}{20} = 0.45$

$\frac{2}{5} = 0.40$

$\frac{11}{25} = 0.44$

$\frac{2}{5}$	$\frac{11}{25}$	$\frac{9}{20}$
Smallest		Largest

2 $\frac{16}{125} = 0.128$

$\frac{31}{250} = 0.124$

$\frac{1}{8} = 0.125$

$\frac{31}{250}$	$\frac{1}{8}$	$\frac{16}{125}$
Smallest		Largest

Percentages (pp. 62–72)

1	Percentage purple	12%
	Percentage white	88%
2	Percentage purple	47%
	Percentage white	53%
3	Percentage purple	20%
	Percentage white	80%
4	Percentage purple	9%
	Percentage white	91%
5	Percentage dark purple	8%
	Percentage light purple	9%
	Percentage white	83%
		Total: 100%
6	Percentage dark purple	13%
	Percentage light purple	15%
	Percentage white	72%
		Total: 100%
7	Percentage light purple	25%
	Fraction light purple	$\frac{25}{100}$
	Fraction light purple	$\frac{1}{4}$
8	Percentage light purple	10%
	Fraction light purple	$\frac{10}{100}$
	Fraction light purple	$\frac{1}{10}$
9	Percentage light purple	20%
	Fraction light purple	$\frac{20}{100}$
	Fraction light purple	$\frac{1}{5}$
10	Percentage light purple	40%
	Fraction light purple	$\frac{40}{100}$
	Fraction light purple	$\frac{2}{5}$

Changing fractions to percentages (p. 64–65)

1 Percentage purple $\frac{3}{10} \times 100 = 30\%$

Percentage white $\frac{7}{10} \times 100 = 70\%$

2 Percentage purple $\frac{4}{20} \times 100 = 20\%$

Percentage white $\frac{16}{20} \times 100 = 80\%$

3 Percentage purple $\frac{10}{25} \times 100 = 40\%$

Percentage white $\frac{15}{25} \times 100 = 60\%$

4 Percentage purple $\frac{6}{25} \times 100 = 24\%$

Percentage white $\frac{19}{25} \times 100 = 76\%$

5 Percentage purple $\frac{18}{50} \times 100 = 36\%$

Percentage white $\frac{32}{50} \times 100 = 64\%$

6 Percentage purple $\frac{23}{50} \times 100 = 46\%$

Percentage white $\frac{27}{50} \times 100 = 54\%$

7 Percentage purple $\frac{7}{20} \times 100 = 35\%$

Percentage white $\frac{13}{20} \times 100 = 65\%$

ISBN: 9780170447331

8 Percentage purple $\frac{17}{20} \times 100 = 85\%$

Percentage white $\frac{3}{20} \times 100 = 15\%$

9	45%	**10**	25%
11	70%	**12**	60%
13	42%	**14**	52%
15	95%	**16**	85%
17	75%	**18**	30%

Changing percentages to fractions (p. 66)

1 $20\% = \frac{20}{100} = \frac{1}{5}$ **2** $75\% = \frac{75}{100} = \frac{3}{4}$

3 $17\% = \frac{17}{100}$ **4** $35\% = \frac{35}{100} = \frac{7}{20}$

5 $90\% = \frac{90}{100} = \frac{9}{10}$ **6** $24\% = \frac{24}{100} = \frac{6}{25}$

7 $62\% = \frac{62}{100} = \frac{31}{50}$ **8** $4\% = \frac{4}{100} = \frac{1}{25}$

9 $\frac{7}{100}$ **10** $\frac{2}{25}$

11 $\frac{12}{25}$ **12** $\frac{17}{20}$

13 $\frac{1}{100}$ **14** 1

Changing decimals to percentages (p. 67)

1	15%	**2**	90%
3	24%	**4**	82%
5	17%	**6**	99%
7	43%	**8**	16%
9	50%	**10**	35%
11	10%	**12**	9%
13	4%	**14**	1%

Changing percentages to decimals (p. 68)

1	0.25	**2**	0.60
3	0.90	**4**	0.17
5	0.85	**6**	0.24
7	0.32	**8**	0.63
9	0.10	**10**	0.72
11	0.05	**12**	0.99
13	1.00	**14**	0.01

Converting between fractions, decimals and percentages (p. 69)

	Fraction	Decimal	Percentage
1	$\frac{1}{2}$	0.5	50%
2	$\frac{3}{10}$	0.3	30%
3	$\frac{2}{5}$	0.4	40%
4	$\frac{1}{4}$	0.25	25%
5	$\frac{19}{20}$	0.95	95%
6	$\frac{4}{5}$	0.8	80%
7	$\frac{1}{5}$	0.2	20%
8	$\frac{3}{4}$	0.75	75%
9	$\frac{1}{10}$	0.1	10%
10	$\frac{10}{10}$	1	100%
11	$\frac{99}{100}$	0.99	99%
12	$\frac{3}{20}$	0.15	15%
13	$\frac{4}{25}$	0.16	16%
14	$\frac{1}{100}$	0.01	1%

Calculating percentages (p. 70)

1	75%	**2**	20%
3	95%	**4**	35%
5	40%	**6**	1%
7	48%	**8**	28%
9	96%	**10**	26.5%

Finding percentages of amounts (pp. 71–72)

1	7	**2**	18
3	150	**4**	520
5	12	**6**	18
7	48	**8**	10 000
9	7.5	**10**	1 188
11	42.9	**12**	512
13	**a** \$16	**b**	\$64
14	**a** \$112.50	**b**	\$337.50

 ISBN: 9780170447331

15 18 hours
16 60
17 0.12 kg

Rounding (pp. 73–80)

Rounding to whole numbers (pp. 73–76)

1 Answer: 50

2 Answer: 670

3 Answer: 1 390

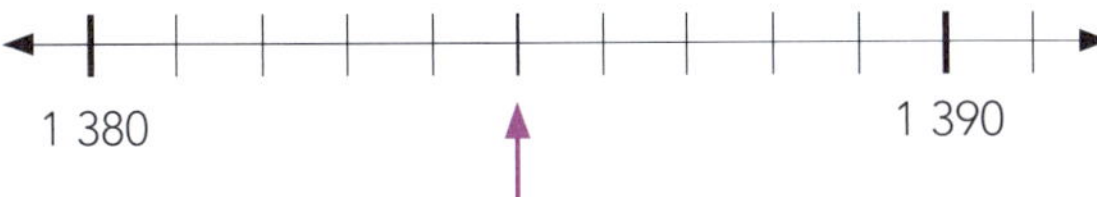

4 Answer: 990

5 Answer: 60 240

6 57 60
7 13 10
8 99 100
9 161 160
10 528 530
11 1 034 1 030
12 6 175 6 180
13 55 682 55 680

14 Answer: 100

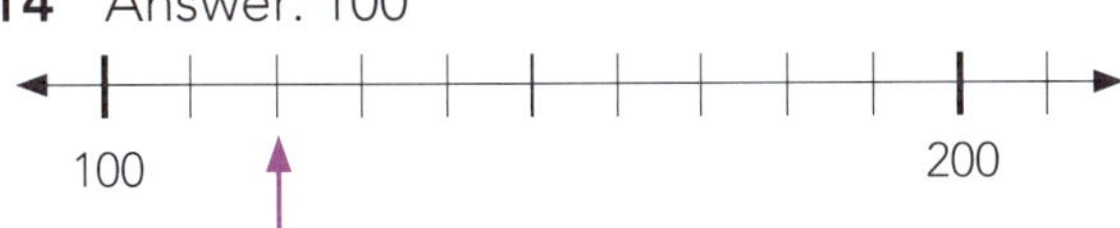

15 Answer: 2 700

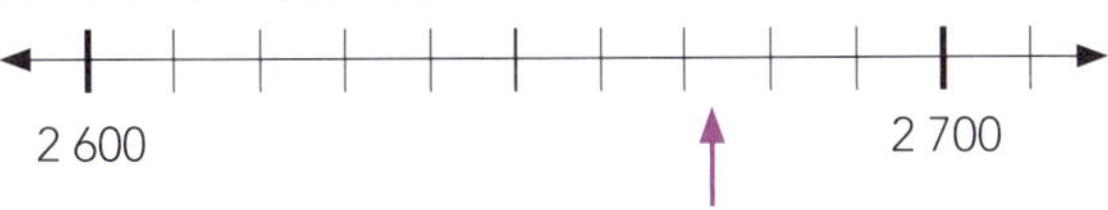

16 Answer: 14 400

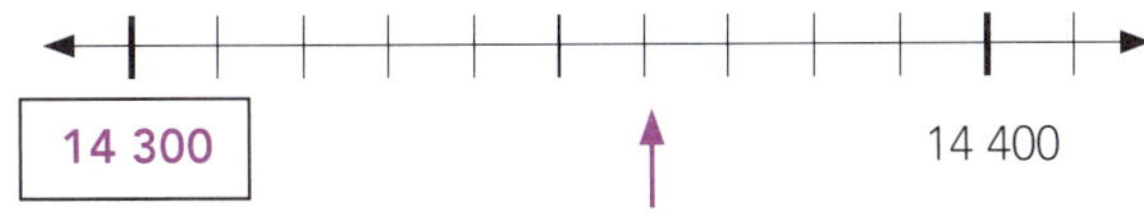

17 Answer: 8 400

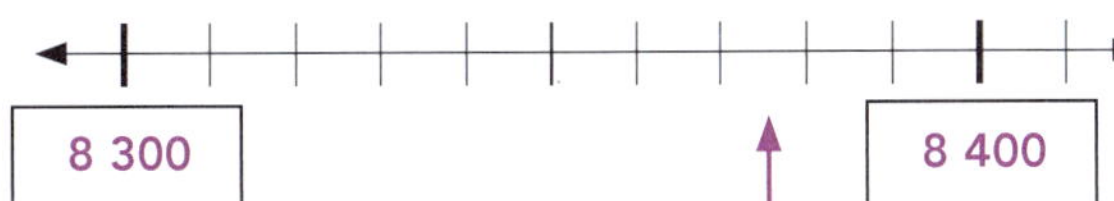

18 Answer: 72 000

19 386 400
20 570 600
21 1 451 1 500
22 6 698 6 700
23 19 674 19 700
24 85 237 85 200
25

Original number	Rounded number	Rounded to the nearest:
135	140	ten
84 918	84 900	hundred
9 451	9 000	thousand
178 679	178 700	hundred
62 455	62 460	ten
40 012 319	40 012 300	hundred

Rounding decimals (pp. 77–80)

1

	Number of decimal places		Number of decimal places
3.568	3	11.09	2
1 324.5	1	372.0	1
412.530	3	0.006	3
0.308	3	5	0
9.4000	4	42.001	3

2 Answer: 23.5

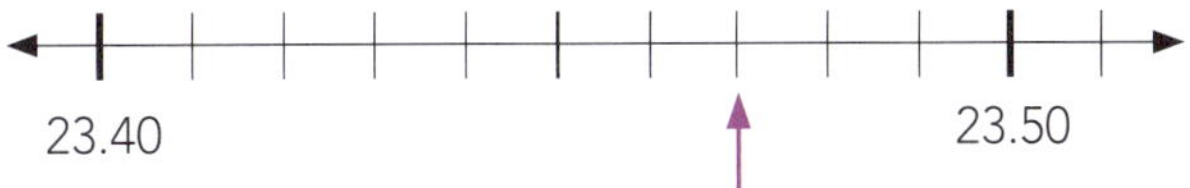

3 Answer: 0.1

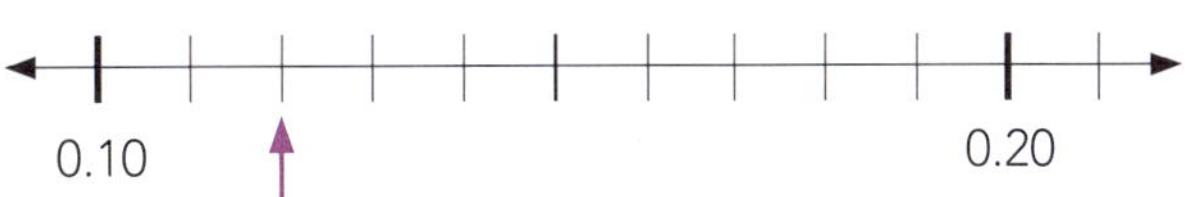

4 Answer: 407.1

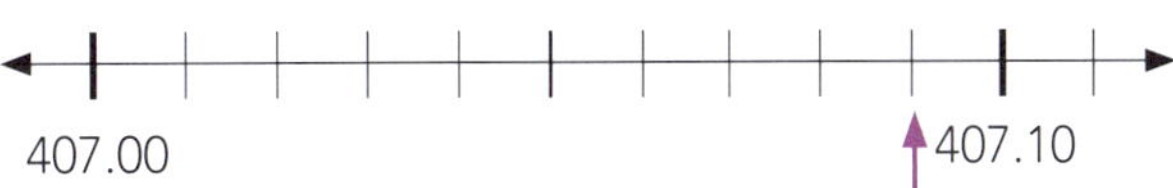

5 Answer: 68.5

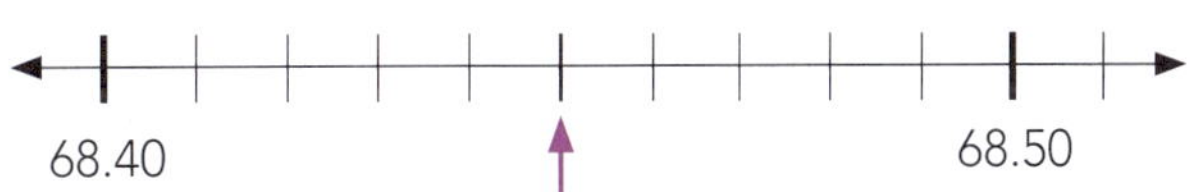

6 Answer: 11.0

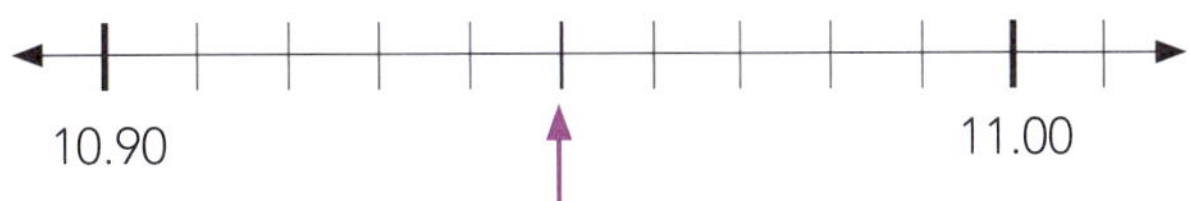

ISBN: 9780170447331

7

	Rounded to 1 dp		Rounded to 1 dp
3.**5**66	**3.6**	11.**6**79	11.7
24.**5**24	24.5	372.**0**9	372.1
2.**9**30	2.9	0.**0**06	0.0
0.**3**08	0.3	5.**9**8	6.0

8 4.2 kg

9 **a** 1.9 kg **b** 7.3 kg

10 **a** 0.7 L **b** 0.5 kg

11 12.9 s

Estimations/approximations (p. 81)

1 9 **2** 11

3 13 **4** 8

5 35 **6** 6

7 30 **8** 10

9 2 **10** 9

11 6 **12** 12

13 22 **14** 10

Revision 1 (pp. 82–84)

1 1 – 3 + 7 – 2 = 3

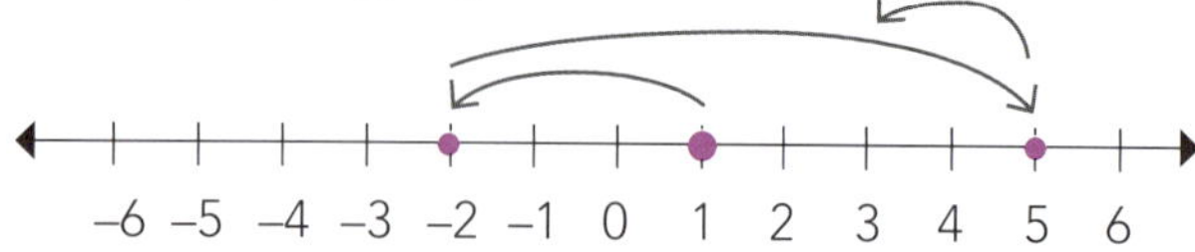

2 **a** 5 **b** –2

3

Multiples				Number	Factors			
12	4	2	5		**1**	4	12	30
30	1	**24**	3	**6**	**2**	**3**	24	36
6	**36**	**18**	28		28	5	18	**6**

4 16

5 **a** 64 **b** 8

6 **a** 5 **b** 10

7 **a** 44 **b** 11

c –2 **d** 23

8 $\frac{2}{5}$ $\frac{3}{7}$

Smaller Larger

9 $\frac{3}{5}$ **10** $\frac{7}{3}$

11 $3\frac{1}{5}$

12 **a** $\frac{3}{20}$ **b** $\frac{4}{7}$

c $\frac{4}{9}$ **d** 21

13

	Number	Numerals	Words
a	5**6**1 328	60 000	Sixty thousand
b	0.3**1**	0.01	One hundredth

14 **a** Fifty thousand and sixty-seven

b Four and eight tenths

15 **a** 902 136

b 0.17

16 **a** 451.7

b 0.73

17

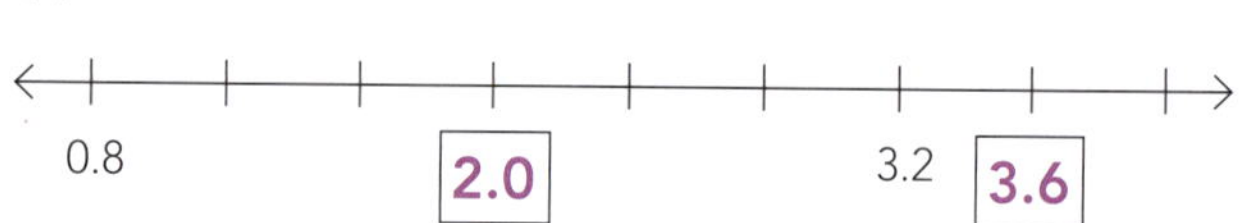

18 **a** 1 040 **b** 0.003

19 Percentage dark purple 13%

Percentage light purple 12%

Percentage white 75%

Total: 100%

20

Fraction	Decimal	Percentage
$\frac{2}{5}$	0.40	40%
$\frac{3}{4}$	0.75	75%
$\frac{3}{20}$	0.15	15%

21 $16.20

22 25%

23

	Round to the nearest:	Highlight the digit you have to round to	Answer
2 387 164	thousand	2 38**7** 164	2 387 000
75 016	ten	75 0**1**6	75 020
62.95	1 dp	62.**9**5	63.0

 ISBN: 9780170447331

Revision 2 (pp. 85–87)

1 $-4 + 6 + 3 - 5 = 0$

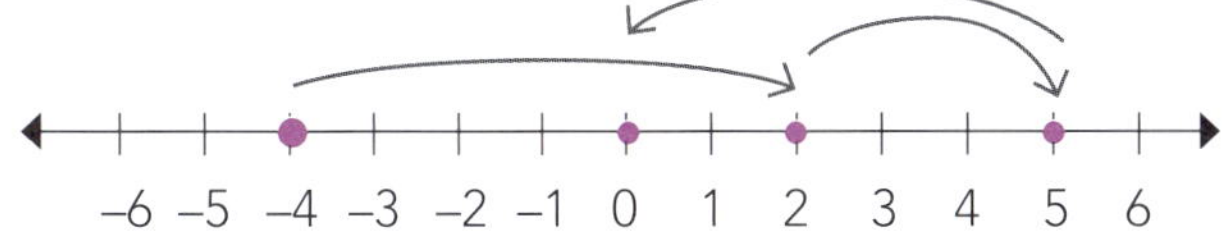

2 a –33 b 13

3

Multiples				Number	Factors			
20	4	30	2		**20**	**4**	30	**2**
10	1	24	**40**	**20**	**10**	**1**	24	40
100	36	5	28		100	36	**5**	28

4 36

5 a 8 b 120

6 a 6 b 3

7 a 38 b 0

c –1 d 99

8 $\frac{2}{3}$ $\frac{7}{10}$

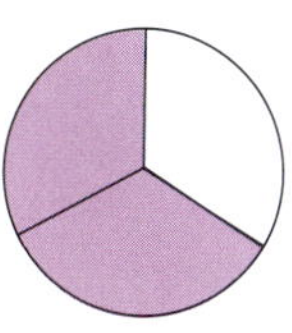
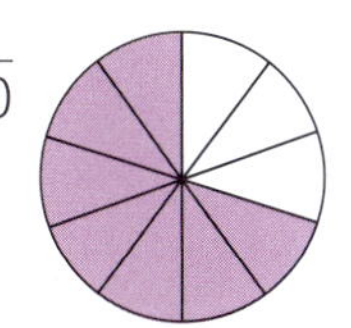

Smaller Larger

9 $\frac{2}{3}$ 10 $\frac{13}{4}$

11 $5\frac{3}{4}$

12 a $\frac{2}{21}$ b $\frac{4}{5}$

c $\frac{4}{12}$ or $\frac{1}{3}$ d 20

13

	Number	Numerals	Words
a	359 400	9000	Nine thousand
b	0.87	0.8	Eight tenths

14 a One hundred and four thousand, eight hundred and four

b Seven and six hundredths

15 a 120 405

b 0.08

16 a 389.712

b 1.31

17

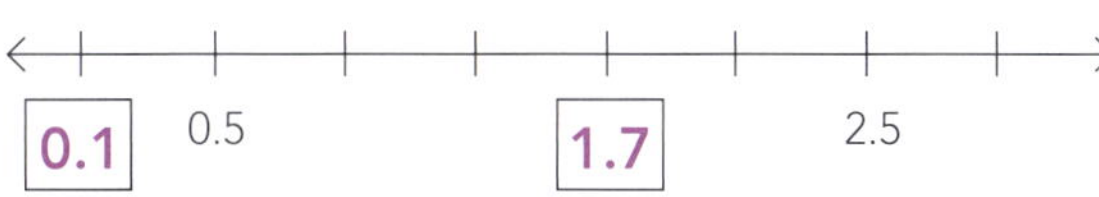

18 a 3 784 b 1.625

19 Percentage dark purple 20%

Percentage light purple 36%

Percentage white 44%

Total: 100%

20

Fraction	Decimal	Percentage
$\frac{7}{10}$	0.70	70%
$\frac{9}{20}$	0.45	45%
$\frac{3}{4}$	0.75	75%

21 $276

22 60%

23

	Round to the nearest:	Highlight the digit you have to round to	Answer
671 892	hundred	671 **8**92	671 900
36.55	whole number	3**6**.55	37
473.07	1 dp	473.**0**7	473.1